Regulation of Agrochemicals

Regulation of Agrochemicals

A Driving Force in Their Evolution

Edited by

Gino J. Marco
Marco Tech

Robert M. Hollingworth
Michigan State University

Jack R. Plimmer
U.S. Department of Agriculture

American Chemical Society, Washington, DC 1991

Library of Congress Cataloging–in–Publication Data

Regulation of agrochemicals: a driving force in their evolution
    Gino J. Marco, editor, Robert M. Hollingworth, editor, Jack R. Plimmer, editor.

        p.   cm.

Based on a symposium honoring the 20th anniversary of the Division of Agrochemicals of the American Chemical Society.

    Includes bibliographical references and index.

    ISBN 0–8412–2089–1 (clothbound) — ISBN 0–8412–2085–9 (paperback)

    1. Agricultural chemicals–Law and legislation–United States–History.
2. Pesticides–Law and legislation–United States–History.

    I. Marco, Gino J., 1924–   . II. Hollingworth, Robert M., 1939–
III. Plimmer, Jack R., 1927–   . IV. American Chemical Society.
Division of Agrochemicals.

KF3959.A75R44   1991
344.73'.04633—dc20   91–19736
[347.3044633]   CIP

1991 ACS Books Advisory Board

Contents

Contributors

Donald G. Crosby   page 9
Department of Environmental Toxicology
College of Agriculture and Environmental Sciences
University of California at Davis
Davis, CA 95616

Barrington Cross   page 89
American Cyanamid Company
P.O. Box 400
Princeton, NJ 08540

Edward C. Gray   page 45
Jellinek, Schwartz, Connally, & Freshman, Inc.
1015 15th Street N.W.
Washington, DC 20005

Ralph W. F. Hardy   page 131
Boyce Thompson Institute for Plant Research
Cornell University
Tower Road
Ithaca, NY 14853–1801

Robert M. Hollingworth   editor
Pesticide Research Center
Michigan State University
East Lansing, MI 48824–1311

G. Wayne Ivie   page 81
Agricultural Research Service
U.S. Department of Agriculture
Research Laboratory
Route 5, Box 810
College Station, TX 77840

Edwin L. Johnson   page 55
Office of International Activities
U.S. EPA (A-106)
401 M Street S.W.
Washington, DC 20460

Philip C. Kearney   page 19
Deputy Area Director
Natural Resources Institute
Room 231, Building 001, BARC-West
Beltsville, MD 20705

J. W. Kobzina   page 121
ICI Agricultural Products
1200 South 47th Street
Box 4023
Richmond, CA 94804-0023

Gustave K. Kohn   pages 1 and 163
198 Pine Lane
Los Altos, CA 94022

Marguerite L. Leng   page 27
Omnitech International
1714 Sylvan Lane
Midland, MI 48640

Keith T. Maddy   page 145
California Department of Food and Agriculture
1220 N Street
Sacramento, CA 95814

Gino J. Marco   editor
Marco Tech
1904 Tennyson Drive
Greensboro, NC 27409

James P. Minyard, Jr.   page 9
Mississippi State Chemical Laboratory
Mississippi State University
Box CR
Mississippi State, MS 39762

Jack R. Plimmer   editor
ABC Laboratories, Inc.
7200 East ABC Lane
P.O. Box 1097
Columbia, MO 65205

James N. Seiber   page 101
Department of Environmental Toxicology
University of California
Davis, CA 95616

Barry Thomas   page 73
Schering Agrochemicals, Ltd.
Chesterford Park Research Station
Saffron Walden, Essex
CB10 1XL England

Preface

REGULATION OF AGROCHEMICALS started in the early part of this century, when agrochemicals were simple inorganic compounds. But those few early state and federal laws had little impact on the growth of the agrochemical industry, which was stimulated by the sometimes desparate need for food, both in America and abroad. For fifty years, the possibility of environmental effects was not considered seriously. By 1962, however, that situation had changed. *Silent Spring* was published, and there was no turning back to virtually unlimited use of any agrochemical product that helped produce good yields. The step-by-step buildup of agrochemical regulation had begun.

Regulations have been a driving force in shaping the efforts of the Agrochemicals Division over the past 20 years. The creation of the Environmental Protection Agency, with its ever-developing regulations, led to responses from the agricultural arena, one of which was that the Pesticide Chemistry subdivision of the ACS Agricultural and Food Chemistry Division became its own division. The change of name from Pesticide Chemistry to Agrochemicals was a further response to directional changes in agrochemicals that were in turn in response to regulations.

Regulations continue to be a driving force in the Agrochemicals Division and the agrochemicals business. Hence, this was the theme of the symposium that honored the 20th anniversary of the Agrochemicals Division and that was the basis of this book.

Acknowledgments

We thank past chairmen Henry Dishburger, Marguerite Leng, and Paul Hedin for acting as co-organizers of the symposium.

GINO J. MARCO
Marco Tech
1904 Tennyson Drive
Greensboro, NC 27409

ROBERT M. HOLLINGWORTH
Pesticide Research Center
Michigan State University
East Lansing, MI 48824–1311

JACK R. PLIMMER
ABC Laboratories, Inc.
P.O. Box 1097
Columbia, MO 65205

Chapter 1

Agrochemicals and the Regulatory Process Before 1970

Gustave K. Kohn

Chemicals used in agriculture, that is, agrochemicals, are not new. For as long as we have farmed the land, we have used chemicals to improve production by killing all kinds of pests. In the early years of the 20th century, the agrochemicals were simple inorganic chemicals, such as lead arsenate and hydrogen cyanide, and naturally occurring organic compounds, such as nicotine and strychnine. These compounds are extremely toxic, and some regulation was necessary to protect the farmer, the shipper, and the consumer (the public at large) from the hazards of such well-recognized compositions.

Early Regulations

By 1900, many states had passed laws that regulated product purity by setting quality control standards. One of the earliest compounds to be so regulated was Paris green (cupric acetoarsenite), an extremely poisonous compound used as an insecticide and as a pigment in paint. In 1910, the Federal Insecticide Act (1) was passed. This act, which was the first specific pesticide regulation, set standards for insecticides and fungicides in interstate commerce. The standards related principally to the quality of the products, that is, the allowable latitude for their chemical compositions. The act regulated the pesticides themselves, not their residues in foods.

2085-9/91/0001$06.00/0 © 1991 American Chemical Society

In the early decades of this century, the United States exported apples in relatively great quantity to Great Britain. Because of numerous reports of consumer poisonings, Great Britain threatened an embargo on the import of American apples. To avoid that consequence, the first true residue tolerance level was established in 1927. Apples could not be shipped in interstate (and intercountry) commerce if they contained more than 3.57 ppm of arsenic residue calculated as As_2O_3 (2). This amount was considered to be the level that the human body could tolerate without ill effects. This regulation is not the first time in the history of the Judeo–Christian and Islamic traditions that the eating of apples changed a way of life! But it is the first time that pesticide residues in foods were regulated. Table I gives a short list of regulations that followed.

In 1938, the Federal Food, Drug, and Cosmetic Act (3) was passed to expand the protection of the public from contamination by chemical residues in foods. This act was the first of a series of laws designed to protect the public. Registering a chemical was tantamount to getting a license to sell it. Registration was granted only after satisfying both the Department of Agriculture that the compound was useful as claimed and the Food and Drug Administration (FDA) that the compound was not a human hazard. The philosophies of these agencies were not identical; in fact, they were at times contradictory. Data came from tests by the manufacturer and by land grant colleges' and universities' plot experiments or similar experiment station studies. Often, data were abstracted from journals like the *Journal of Economic Entomology*.

Environmental considerations were minimal, although the Food and Drug Administration required information on potential human health hazards. Short-, intermediate-, and long-term animal studies were required for establishment of safety.

The Organic Pesticide Revolution and Its Effects

A great revolution took place at the time of World War II when the useful biological effects of the extraordinarily potent, allegedly innocuous synthetic organic pesticides, particularly DDT and 2,4-D, were discovered. After all, American agriculture lit-

Table I. Federal Regulations Enacted Before 1970

Year	Regulation	Content	Ref.
1910	Federal Insecticide Act	First specific pesticide regulation; set standards for insecticides and fungicides moving in interstate commerce	1
1938	Federal Food, Drug, and Cosmetic Act	Protected the public from contamination of food, drugs, and cosmetics	3
1947	Federal Insecticide, Fungicide, and Rodenticide Act (FIFRA)	Regulated interstate shipment of pesticides; set safety standards for handling of chemicals in interstate commerce; required Federal registration before sale	6
1954	Miller Pesticide Amendment to FIFRA (PL 518)	Required setting of residue tolerances for pesticides on raw agricultural products	7
1958	Food Additives Amendment (contains Delaney Amendment)	Protected public against contamination of processed agricultural products and foods by pesticide and other contaminants; prohibited direct food additives that cause cancer at any dosage level of test protocol	8
1970	Environmental Protection Agency (EPA) established by Executive fiat	EPA assumed regulatory responsibility for pesticides; proof of environmental and consumer safety required for registration	10

erally was feeding much of the world after the devastation of most of Europe and parts of Asia and Africa. A key aspect of American foreign policy was to provide food for all of the non-Communist world. Consequently, an immense quantitative increase in the production and use of pesticidal chemicals occurred. After 1945, the production of synthetic organic pesticides increased so dramatically that inorganic pesticides dropped from more than 75% to less than 10% of total sales, and synthetic organic pesticides comprised more than 90% of total sales (4). Production of synthetic organic insecticides and herbicides steadily increased from less than 50 million pounds in the early 1940s to more than 400 million pounds in the 1970s (5).

The Federal Insecticide, Fungicide, and Rodenticide Act (6) (FIFRA) was passed in 1947. FIFRA regulated the interstate shipment of pesticides and set safety standards for the handling of chemicals in interstate commerce.

FIFRA was frequently amended. In 1954, the Miller Pesticide Amendment (7) required the setting of residue tolerance levels for pesticides on raw agricultural products. The 1958 Food Additives Amendment, which contains the Delaney Amendment (8), expanded protection by setting tolerance levels for pesticides on processed agricultural products and foods. The Delaney Amendment established zero tolerance levels for any chemical in food in which animal tests at any level exhibited carcinogenicity. A zero residue tolerance level means that the compound cannot be detected by the best practical analytical method available. Between 1947 and 1966, most states passed laws extending FIFRA and Federal food additives legislation to supplement protection within state boundaries.

Industry responded to FIFRA by greatly increasing its research and development efforts. These efforts included exploration of analytical methodologies and practices that could accurately determine residues of pesticides and their metabolites in complex substrates such as plant and animal tissues, foods, and soils. Eventually, such residues and metabolites could be determined to the heretofore unknown levels of parts per million (ppm) and parts per billion (ppb).

FIFRA stimulated further improvements in process research, primarily to minimize toxic impurities and reduce costs; safer pest-controlling compositions were made more easily available. At the same time the discovery and development of new useful compositions with lessened human toxicity was encouraged and achieved. FIFRA, together with U.S. political and economic objectives, stimulated the American agrochemical industry to expand and achieve a dominant position in the world in the 1960s and 1970s.

At first, environmental effects and the probability for organism resistance were not considered. Chemists and entrepreneurs for the most part ignored organism adaptation, but they were not alone. Even the biologists did, as a reading of the *Journal of Economic Entomology* during most of the late 1940s and early 1950s attests. However, evidence of persistence in the environment, environmental metabolism (including intoxication), and organism resistance was mounting. *Silent Spring* was published in 1962 (*9*). The establishment of the Environmental Protection Agency (EPA) was inevitable. In 1970, the EPA was established by executive order (*10*). EPA assumed regulatory responsibility for pesticides. Proof of environmental safety and consumer safety was required before a compound could be registered.

Chemistry and the Chemist Before the Establishment of the EPA

Although a botanist had discovered chromatography (*11*), none of the common forms of this separation technique was practiced (or even known) in the chemical laboratories of the 1940s. We speak glibly today of part-per-billion residues and solution concentrations of 10^{-15} and even 10^{-18} M, but in the 1940s, such low levels were not measurable. Analysis was, for the most part, wet gravimetric or volumetric chemistry. For example, arsenic compounds (like many inorganic compounds) were analyzed by a volumetric iodometric method that, at best, could yield values with three to four significant figures. Nicotine, a natural aphicide, was analyzed that way. The substrate was extracted, and the nicotine was precipitated from solution as a complex silicotungstate salt. The salt was filtered, and the precipitate was ignited, dried, and weighed. I remember using a small, bent-glass rod attached to a stirrer and fitted to scratch the inner surface of a beaker to hasten crystallization and provide macroscopic crystals for rapid separation by filtration. There were no recording automatic balances, and the fourth, sometimes the fifth, decimal place was estimated by the counting of amplitudes of successive swings of the balance pendulum. Believe it or not, these results were reproducible and reasonably accurate.

For the most part, particularly in the late 1940s and early 1950s, except with enzymatic analyses and radiochemistry, accurate part-per-million and part-per-billion determinations were unobtainable. We learned at an early date how sensitive enzymatic analytical methods could be. Our laboratory was able to determine the residues of the pesticide tetraethylpyrophosphate (TEPP) (used in dairies for fly control) in milk by enzymatic analysis. TEPP is hydrolytically very unstable and breaks down into various phosphate entities and ethanol. These fragments are ubiquitous in plant tissue, and ordinary noninstrumental analysis (e.g., a gravimetric phosphate determination) would require subtracting very small differences from very relatively large numbers; hence this conventional approach is analytically unacceptable. Our laboratory, working cooperatively with the University of California at Berkeley, purchased a very sensitive potentiometer with which we satisfied the FDA, through pH determination based on enzymatic chemistry, that there was a zero residue tolerance level in milk if spraying was conducted by an agreed-upon protocol.

At one point, my employer was involved in a large lawsuit for which it was claimed that a rather small sulfur contamination of TEPP resulted in the destruction of a large tract of land that was growing vegetables (mostly cucumbers). We sent a sample of KCl to Oak Ridge where it was bombarded with neutrons and returned to us containing a soft β-ray emitting isotope of sulfur with a usable half-life. We sprayed cucumber plants with TEPP containing the sulfur isotope in a variable ratio. This same employer at that time ran the Livermore atomic energy facility. I remember being handed over a barbed wire fence (I had no clearance) two instruments that were then state-of-the-art Geiger counters. Using these Geiger counters, we followed the residues of sulfur on the plants in the greenhouse up to harvest and in the various formulation mixes used in the sprays. The results indicated that sulfur in TEPP caused no injury below an order of magnitude higher residue than that was reported by the opposing side.

Modern instrumentation and technology were introduced at various times after 1950 and more after 1970. IR instruments for routine analysis were not available until well into the 1950s.

The agrochemical industry was an infant in the late 1940s. The companies, by today's standards, were small. A chemist could provide an idea, synthesize the molecule, work on the manufacturing process and on the start-up of the manufacturing plant, create a usable formulation, determine metabolism and degradation, devise analytical procedures, and so on. Today, in the mammoth organizations that now exist, the individual chemist barely knows the names of his colleagues working on the various complex aspects of the development of a new agrochemical. This situation is the inevitable result of the maturation of an industry and the consolidation of companies. In the 1940s, 1950s, and 1960s, agrochemicals was a new growth industry. It resembled in this respect semiconductors and computers in the 1970s and biotechnology in the early 1980s.

References

1. "Federal Insecticide Act", Public Law 152, *U.S. Statutes at Large* 1, pp 331–335; April 26, 1910.
2. Food and Drug Regulation, 1927.
3. "Federal Food, Drug, and Cosmetic Act", Public Law 95–532, *U.S.C.A.* Title 21, 321 A,B,C.
4. Based on data from Chevron Chemical Company.
5. Kohn, G. K. "The Pesticide Industry" in *Riegel's Handbook of Industrial Chemistry*; Van Nostrand: New York, 1983; p 750.
6. "Federal Insecticide, Fungicide, and Rodenticide Act", Public Law 92–516, *U.S. Code* Title 7, Pt. 136 et seq., October 21, 1972.
7. Miller Amendment, Public Law 518, *U.S. Statutes at Large* 68, p 511, July 22, 1954.
8. "Food Additives Amendment to FIFRA", Public Law 85–929, *U.S. Statutes at Large* 72, p 1784, September 6, 1958.
9. Carson, Rachel *Silent Spring*; Houghton Mifflin: Boston, 1962.
10. *Code of Federal Regulations* Title 3, 1966–1970 comp.
11. Swett, T. *Ber. Dtsch. Bot. Ges.* **1906**.

Chapter 2

The Persistent Seventies

Donald G. Crosby and James P. Minyard, Jr.

The 1970s formed a pivotal decade for pesticide science—perhaps the most important decade. The age of the chlorinated hydrocarbon insecticides was drawing to a close in much of the world, and the last years of the 1960s had seen an awakening of activity in all major aspects of pesticide science. In accordance with this trend, the American Chemical Society's Division of Pesticide Chemistry was formed in 1969 to provide a forum for academic, government, and industrial scientists. The term "pesticide" became generic for all types of pest-control chemicals.

The environmental movement also was in full swing, and the first Whole Earth Day was celebrated on May 1, 1970. Persistent pesticides were one of the principal concerns of environmentalists, and an increasingly vocal public was demanding stricter regulation of agricultural chemicals and more environmentally compatible alternatives. Indeed, the agricultural use of pesticide chemicals increased rapidly, and the total amount applied more than doubled between 1970 and 1980 (*1*) (Figure 1). The U.S. Department of Agriculture (USDA) was perceived, rightly or wrongly, as being too close to the beneficial uses of pesticides to be fully effective in regulating them.

The U.S. Environmental Protection Agency (EPA) was formed in December 1970 by President Nixon's Reorganization Order Number 3, and the agency established an Office of Pesticide Programs to assume and expand the USDA's pesticide regulatory functions. This was the first of many major events in

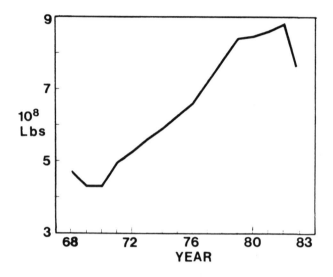

Figure 1. Agricultural use of pesticides in the United States, 1968–1983. (Adapted from reference 1.)

pesticide science and regulation in the 1970s (Table I). In particular, the data requirements for registration of new pesticides and reregistration of existing ones were enlarged, and the 1968 guidelines proposed by the Food and Drug Administration (FDA) were reintroduced by EPA in June 1975, in a greatly enlarged and more scientifically detailed form (2). These guidelines, still basically in force today, were especially significant because they provided the impetus for much of the pesticide research and testing to follow, not only in the United States, but throughout the world. Inherent in them was the pursuit of improved safety and decreased environmental persistence.

The Federal Insecticide, Fungicide, and Rodenticide Act (FIFRA) was amended in 1972 by the Federal Environmental Pesticide Control Act (FEPCA) (and amended again in 1975 and 1978) to reflect the rapid changes in the perception of pesticide chemicals caused by improved analytical capability, health perspectives, and modern application techniques. FEPCA provided a major stimulus to the search for a balance between pesticide efficacy and environmental protection by essentially rewriting FIFRA to provide EPA with more flexibility, stronger enforcement, and a shift of legal emphasis from efficacy toward health and environment.

Table I. Some Major Events in Pesticide Science and Regulation in the 1970s

Year	Event
1970	Environmental Protection Agency (EPA) established.
	National Institute for Occupational Safety and Health (NIOSH) established.
1971	ACS Chlorodioxin Symposium.
	Glyphosate herbicide introduced.
1972	Federal Environmental Pesticide Control Act (FEPCA) enacted.
	All major uses of DDT suspended.
	First microencapsulated pesticide introduced (methylparathion).
1973	Synthetic insect growth regulator diflubenzuron patented.
	Insect growth regulating properties of methoprene described.
	First table of accurate pesticide vapor pressures published.
1974	Third International Congress of Pesticide Chemistry (Helsinki).
	First standards set for work re-entry into pesticide treated fields (EPA).
1975	EPA Guidelines for Registering Pesticides in the United States.
	Rebuttable Presumption Against Registration (RPAR) established.
	Pesticide Science Society of Japan organized.
	First insect growth regulator registered with EPA (methoprene).
1976	Federal Resource Conservation and Recovery Act (RCRA) enacted.
	RPAR issued for strychnine, endrin, Kepone, 1080, and BHC.
	Most pesticidal uses of mercury compounds canceled (EPA).
1977	Federal Toxic Substances Control Act (TSCA) enacted.
	EPA and USDA embrace concept of integrated pest management (IPM).
1978	Sulfonylurea herbicides patented.
	Fourth International Congress of Pesticide Chemistry (Zürich).
	EPA concludes first full-scale RPAR (chlorbenzilate).
	First list of restricted-use pesticides published by EPA.
1979	All major uses of 2,4,5-T and silvex suspended.
	First report of avermectins.
	First registration of a pheromone (gossyplure for pink bollworm).

Developing Less Persistent Pesticides

This is not to say that concern for the environmental and health aspects of pesticides had been lacking previously. The development and improvement of less persistent carbamate and organophosphorus insecticides had begun in the 1960s, and the products of this major effort had started to appear on the commercial market. Two especially important examples were the simple, safe phosphoramidate, acephate (Orthene), introduced in 1970, and the first commercially successful synthetic pyrethroid, bioresmethrin, introduced in 1969.

Indeed, the 1970s brought renewed interest in natural pest control agents (*3*), and the term "biorational" was introduced to describe pesticide structures based on natural products. Other synthetic pyrethroids reached the market, including permethrin and deltamethrin in 1973, and showed that, unlike the natural pyrethrins, almost any degree of persistence could be designed into such molecules (*4*). Bacterial and viral insecticides, such as those from *Bacillus thuringiensis* and *Heliothus virescens*, became increasingly promising, and a *B. thuringiensis* formulation (Thuricide) finally was registered after a long struggle. The search for new commercially practical natural pesticides, which had slowed in the previous decade, was stimulated by development of cartap, a simple sulfur-containing insecticide related to a secretion from an annelid worm (*5*), and climaxed by the discovery of a spectacular group of highly polar microbial biocides, the avermectins (*6*). Insect growth regulators such as methoprene (*7*) and pheromones such as the Gypsy moth attractant, disparlure (*8*), also achieved practicality.

Herbicides, too, had been under active development in the 1960s and, in the 1970s, surpassed the combined totals of all other types of pesticides in both volume of use and economic value (*1*) (Table II). From the comparatively few structural classes of the previous decade, over a dozen new herbicide types emerged as commercially important, among them the diphenyl ethers such as diclofop, acylanilides such as butachlor, and unusual *N*-heterocyclic compounds such as hexazinone. In 1978, a major new class of herbicide, the sulfonylureas, was patented

Table II. Pesticides Applied to Major U.S. Crops, 1966–1982

Year	Herbicides	Insecticides	Fungicides	Other	Total
1966	101.2	108.3	6.0	35.7	251.1
1971	213.1	127.9	6.4	29.8	377.2
1976	373.9	130.3	8.1	35.3	547.6
1982	455.6	71.2	6.6	24.3	557.7

NOTES: Major crops are alfalfa and other hay, corn, cotton, pasture, peanuts, rice, sorghum, soybeans, tobacco, wheat, and other small grains.
All values are given in millions of pounds of active ingredient, excluding sulfur and petroleum.
SOURCE: Reference 1.

(*9*), and some 230 patents on this class were filed during the following decade.

However, as more research was devoted to them, it became apparent that many herbicides considered to be biodegradable and short-lived were much more persistent than had been previously assumed. In particular, phenoxy herbicides such as 2,4,5-T [(2,4,5-trichlorophenoxy)acetic acid] and silvex were found to contain persistent and highly toxic impurities, the chlorinated dibenzodioxins and dibenzofurans, and the term Agent Orange (referring to a 2,4,5-T–2,4-D [(2,4-dichlorophenoxy)acetic acid] ester mixture used in the Vietnam War) became synonymous with "herbicide" in the minds of many people. Furthermore, dinitroanilines and other amine herbicides were discovered to be contaminated with toxic and carcinogenic *N*-nitrosamines, and a carcinogenic byproduct, ETU (ethylenethiourea), was found in the widely used dithiocarbamate fungicides.

Concerns of the Scientific Community

The 1971 pioneering ACS symposium on chlorinated dioxins (*10*), followed by the 1972 symposium on ETU (*11*) and another on *N*-nitrosamines in 1978 (*12*), highlighted the concerns of both the scientific community and the public and led to strict regulation of nitrosamine content, continuing regulatory suspicion of the dithiocarbamates, and complete suspension of the registrations of dioxin-containing herbicides based on 2,4,5-trichlorophenol.

In the 1970s, pesticide scientists and regulators became acutely aware of environmental chemodynamics, the physical movement and chemical fate of pesticides (*13*). This concept included the abiotic processes of volatilization, adsorption, hydrolysis, and photochemical degradation as well as chemical and biochemical degradation. Because of the then immeasurably low vapor pressures of most pesticides, volatility had not usually been considered to be important; however, introduction of Spencer and Cliath's sensitive gas chromatographic method for vapor pressure measurement (*14*), and the theoretical and practical work of Mackay (*15*) on Henry's law, showed volatility to be a major force in pesticide movement through the environment. Pesticide photodegradation in water (16), in air (17), and on surfaces (18) was explored and found to involve primarily oxidation, reduction, and nucleophilic displacement reactions.

Bioconcentration, originally considered to be a highly complex process, was shown as basically only solvent partitioning (19–21), and this finding opened the way for straightforward mathematical modeling and prediction of bioconcentration factors (22). In fact, all movement of pesticides between air, water, soil, and biota soon was recognized as conforming to relatively simple partition principles (23), and regulatory agencies began to require applicants to include standardized measures of necessary physical constants, photolysis, and biodegradation data for registration and reregistration of pesticides in relation to environmental persistence.

Regulatory Action in the 1970s

As the decade progressed, increasing scientific and regulatory attention was paid to applicator and farm-worker safety (24). In 1971, the recently formed National Institute for Occupational Safety and Health (NIOSH) became interested in the protection of workers from pesticides, and, although the results became controversial, EPA issued its first Proposed Occupational Safety and Health Standards for Farm Workers Dealing with Pesticides in 1974 (25). These worker safety intervals ("reentry times") were similar to those that had been put in place in California in 1971. On the practical side, a number of protective techniques were introduced (26), including reduction of spray drift through thickeners, better protective clothing, and improved warning signs in both English and Spanish. In addition, new compounds such as the herbicide triclopyr (Garlon) and the insect growth regulator diflubenzuron (Dimilin), introduced during this decade, often required applications of only small fractions of a kilogram per hectare to be effective.

Within several years after its formation, EPA introduced the RPAR (Rebuttable Presumption Against Reregistration) process, later known as Special Review, to investigate and challenge the continued use of certain pesticides whose toxic or environmental properties were of particular concern. Although few registrations actually were canceled or suspended (1), the process did generate valuable discussion and information and provided ongoing means for monitoring and improving pesticide safety.

Pesticide waste and container disposal also received widespread attention (27, 28). Chemical methods (hydrolysis, oxidation, and chlorinolysis), physical methods (incineration and charcoal adsorption), and biological methods (microbial hydrolysis, microbial oxidation, and soil incorporation) came under active research and initial practical application. EPA issued guidelines for disposal of pesticide wastes and containers in 1974 (29), and the Federal Resource Conservation and Recovery Act (RCRA) (1976) and Toxic Substances Control Act (TSCA) (1977) also influenced the development and implementation of safer disposal methods.

Although not new in concept, the practical application of integrated pest management (IPM) finally was realized during the 1970s (30). IPM combines the use of pesticides with cultural and biological controls. An executive order of the President of the United States established a multidisciplinary research effort among 19 universities and several government agencies known as the Huffaker IPM project (formally, the Principles, Strategies, and Tactics of Pest Population Regulation and Control in Major Crop Ecosystems) which continued for a period of years with funding of more than $1 million per year. IPM is still finding increasing application today. For this reason, together with changing economics, concern over pest resistance, and improved materials and formulations, pesticide use never again attained the almost 1-billion-pound annual level reached in 1982 (1).

* * *

The 1970s was a key period in the history of pesticide discovery, application, and regulation. By the decade's close, the most persistent chemicals had been phased out over much of the world, the present body of pesticide law was largely in place, and worldwide recognition had been given to the need for environmental and human protection as pesticide production and use matured. The persisting groundwork for pesticide science and regulation had been laid for the 1980s and beyond.

References

1. Osteen, C. D.; Szmedra, P. I. *Agricultural Pesticide Use: Trends and Policy Issues*; Economic Research Service, U.S. Department of Agriculture: Washington, DC, 1989; Agricultural Economic Report No. 622.
2. *Federal Register* 40 (1975) pp 26802–26830.
3. Jacobson, M.; Crosby, D. G. *Naturally Occurring Insecticides*; Marcel Dekker: New York, 1972.
4. *Synthetic Pyrethroids*; Elliott, M., Ed.; ACS Symposium Series 42; American Chemical Society: Washington, DC, 1977.
5. Konishi, K. *Agric. Biol. Chem.* **1970,** *34,* 935–940.
6. Miller, T. W.; Chaiet, L.; Cole, D. J.; Cole, L. J.; Flor, J. E.; Goegelman, R. T.; Gullo, V. P.; Kempf, A. J.; Krellwitz, W. R.; Monaghan, R. L.; Ormond, R. E.; Wilson, K. E.; Albers-Schonberg, G.; Putter, I. *Antimicrob. Agents Chemother.* **1979,** *15,* 368.
7. Henrick, C. A.; Staal, G. B.; Siddal, J. B. *J. Agric. Food Chem.* **1973,** *21,* 354–359.
8. Beroza, M. In *Pest Management with Insect Sex Attractants*; Beroza, M., Ed.; ACS Symposium Series 23; American Chemical Society: Washington, DC, 1976.
9. Beyer, E. M., Jr.; Duffy, M. J.; Hay, J. V.; Schlueter, D. In *Herbicides: Chemistry, Degradation, and Mode of Action*; Kearney, P. C.; Kaufman, D. D., Eds.; Marcel Dekker: New York, 1988; pp 117–189.
10. *Chlorodioxins: Origin and Fate*; Advances in Chemistry Series 120; Blair, E. H., Ed.; American Chemical Society: Washington, DC, 1973.
11. "Symposium on Origin and Fate of Ethylenethiourea Fungicides"; Tweedy, B. G., Ed. *J. Agric. Food Chem.* **1973,** *21,* 323.
12. *N-Nitrosamines*; Anselme, J.-P., Ed.; ACS Symposium Series 101; American Chemical Society: Washington, DC, 1979.
13. Haque, R.; Freed, V. H. *Residue Rev.* **1974,** *52,* 89–116.
14. Spencer, W. F.; Cliath, M. M. *Environ. Sci. Technol.* **1969,** *3,* 670–674.
15. Mackay, D.; Leinonen, P. J. *Environ. Sci. Technol.* **1975,** *9,* 1178–1180.
16. Crosby, D. G. In *Fate of Organic Pesticides in the Aquatic Environment*; Faust, S. D., Ed.; Advances in Chemistry Series 111; American Chemical Society: Washington, DC, 1972; pp 173–188.

17. Crosby, D. G.; Moilanen, K. W. *Arch. Environ. Contam. Toxicol.* **1974**, *2*, 62.
18. Lykken, L. In *Environmental Toxicology of Pesticides*; Matsumura, F.; Boush, G. M.; Misato, T., Eds.; Academic: New York, 1972; pp 449–469.
19. Hamelink, J. L.; Waybrant, R. C.; Ball, R. C. *Trans. Am. Fish. Soc.* **1971**, *100*, 207.
20. Hamelink, J. L. In *Aquatic Toxicology and Hazard Evaluation*; Mayer, F. L.; Hamelink, J. L., Eds.; ASTM: Philadelphia, PA, 1977; pp 149–161.
21. Crosby, D. G. *Pure Appl. Chem.* **1975**, *42*, 233–253.
22. Neely, W. B.; Branson, D. R.; Blau, G. E. *Environ. Sci. Technol.* **1974**, *8*, 1113–1115.
23. Lyman, W. J.; Reehl, W. F.; Rosenblatt, D. H. *Handbook of Chemical Property Estimation Methods*; McGraw–Hill: New York, 1982 (original U.S. Army report, 1979).
24. Gunther, F. A.; Gunther, J. D. *Residue Rev.* **1976**, *62*, 1–174.
25. *Federal Register* 39 (1974) pp 9457–9462.
26. Bailey, J. B. *Agric. Age* **1972**, *28*, 6.
27. Munnecke, D. M. *Residue Rev.* **1979**, *70*, 1–26.
28. *Disposal and Decontamination of Pesticides*; Kennedy, M. V., Ed.; ACS Symposium Series 73, American Chemical Society: Washington, DC, 1978.
29. *Federal Register* 39 (1974) p 15236.
30. *Integrated Pest Management*; Apple, J. L.; Smith, R. F., Eds.; Plenum: New York, 1976.

Chapter 3

Impact of Regulations on the ACS Division of Pesticide Chemistry

Philip C. Kearney

The history of the development of the pesticide registration process by the U.S. Environmental Protection Agency (EPA) and the chemical principles developed during the formative years of the ACS Division of Pesticide Chemistry that gave credence to the registration process bear many parallels. The Division of Pesticide Chemistry (now the Division of Agrochemicals) was established in 1969, and the administration of the Federal Insecticide, Fungicide, and Rodenticide Act (FIFRA) was transferred from the U.S. Department of Agriculture (USDA) to the EPA in 1970. At that time, there was an urgent need to develop a scientifically sound set of registration guidelines that were meaningful from an environmental standpoint and legally defensible from an EPA standpoint. The newly formed EPA was very concerned that unless their regulatory posture was scientifically acceptable, their policy decisions would be open to serious debate.

The objective of this review is to trace the history of pesticide chemistry over the past 20 years and the impact of those findings on the regulatory process as we understand it today. It will primarily address the environmental chemistry research aspects of pesticides that influenced the registration process.

Early Guidelines

One of the first major documents that had an impact on the regulation of pesticides and research was the "Report of the Secretary's Commission on Pesticides and Their Relationship to Environmental Health" (1) in 1969. That early document provided a comprehensive literature review and an assessment of the environmental and human health implications of a number of pesticides. It also spelled out some of the environmental concerns about pesticides and identified the need for a stronger regulatory program.

The first environmental guidelines were probably contained in PR Notice 70–15 (2) from the Pesticides Regulatory Division, which was still part of the Agricultural Research Service in 1970. Those guidelines included six proposed studies that really form the backbone for environmental studies as known today:

1. What is the rate of dissipation of the pesticide in soils?

2. What is the mechanism of degradation of the pesticide residues? This study had four subparts:

   - photodecomposition
   - the effect of metabolism on the residues
   - the effect of the residues on microorganisms
   - degradation in water

3. Do residues leach in soils?

4. Do residues move in surface water?

5. Is the pesticide bound, and is it active in the bound state?

6. What levels will accumulate in fish, rabbit, and bird tissues, and what dose-related symptoms are exhibited?

By today's standards, these requirements seem obvious, but they appeared when the Division of Pesticide Chemistry was 1 year old and before EPA was formed. There were no set standards for conducting these studies and, in some cases, no tested methodology.

The early 1970s were exciting times in pesticide chemistry in method development. Every new advance in pesticide environmental chemistry required some innovative techniques. Most of those methods came from members of the Division of Pesticide Chemistry. The only detriment to the present system is that the strict adherence to prescribed guidelines has distracted researchers from conducting innovative research still needed in the regulatory process.

Role of the Division of Pesticide Chemistry in Regulatory Guidelines

The Division of Pesticide Chemistry moved rapidly to fill some of the gaps needed in developing a scientifically reproducible and sound basis for regulating pesticides. Prior to the formation of the division, a symposium was sponsored on "Organic Pesticides in the Environment" in 1966, under the auspices of the Pesticide Subdivision of the Division of Agricultural and Food Chemistry.

At its first meeting in New York in 1969, the Division of Pesticide Chemistry sponsored symposia on "Photodecomposition of Pesticides" and on "Sediment–Water Interactions of Pesticides". Subsequently, the division sponsored symposia on such topics as:

- "Identification of Pesticides at the Submicrogram Level"
- "Pesticide Regulation in the United States and Canada"
- "Fate of Pesticides in Aquatic Environments"
- "Environmental Dynamics of Pesticides"
- "Bound and Conjugated Pesticide Residues"
- "Guidelines for Environmental Studies"
- "Pesticide Analytical Methodology"
- "Pesticides, Policies and Regulations"
- "Pesticide Chemistry in the Twentieth Century"
- "Pesticides in the Air"
- "Fate of Pesticides in Large Animals"
- "Tests for Pesticide Environmental Behavior"

The division also sponsored a special conference on "Bound Residues" in Vail, Colorado. These symposia and general sessions papers provided a wealth of ideas and methods for EPA to develop regulatory guidelines. They addressed all the issues raised in PR Notice 70–15.

What became clear from all of these activities was that some techniques could be transferred from classical chemistry, and some new techniques would be required. A few examples can be cited to illustrate the role Division chemists played in chemical technology transfer.

Detection.   One of the greatest contributions that pesticide chemists have made to environmental chemistry was the development of analytical methodology for the detection of very dilute concentrations of residues in a variety of environmental substrates. Much of the regulatory language prior to the enactment of FIFRA was based on zero residue levels detected in agricultural commodities. The advent of gas chromatography and the transfer of this technology to pesticide residue analysis allowed regulations to be established on actual pesticide concentrations. The original discovery by James and Martin (*3*) of a simple gas chromatograph was rapidly transferred to residue analysis.

The symposium on identification of pesticides at the submicrogram level served as an excellent summary of the state of art at that time. Moye's book, *Analysis of Pesticide Residues* (*4*), traces the individual contributions of pesticide chemists in the development of specific columns and detectors. An excellent chapter in Moye's book (*5*) deals with Federal requirements for pesticide residue analysis and documents the increasing demand for and use of pesticide data in the regulatory process.

Hydrolysis.   Hydrolysis studies were fairly easy to transfer to pesticide solutions. These studies are used to establish the significance of chemical hydrolysis as a route of degradation and to identify the products that might adversely affect nontarget organisms or accumulate in food chains. Conditions for hydrolysis studies are well spelled out with regard to concentration, temperature (25 °C), and pH (5, 7, and 9). This pH range would cover most soil and water situations. Important contributions to the establishment of guidelines were made by Krzeminski et al. (*6*) on the effect of temperature and pH on rates of hydrolysis and Gomaa et al. (*7*) on the kinetics and mechanisms of acid–base-catalyzed hydrolysis of organophosphates.

Photolysis. Techniques for photochemical studies were fairly well worked out in classical organic chemistry, but their transfer to environmental samples offered some challenging problems. The pesticide photochemist was forced to work with dilute aqueous solutions that were often contaminated with other substances. An excellent early review of pesticide photochemistry by Crosby and Li (*8*) discussed the principles of herbicide photochemistry, equipment, and reactions of nine major classes of herbicides.

Unfortunately, most of the early photolysis experiments were conducted in the laboratory under artificial conditions, generally in organic solvents and using mercury lamps emitting principally at 254 μm. There was limited evidence of photolysis under actual field conditions, and only two examples could be cited. Slade (*9*) was able to demonstrate the photolysis of paraquat to 1-methylpyridinium-4-carboxylate on leaves of several crop plants. Kuwahara et al. (*10*) demonstrated that pentachlorophenol was photodecomposed in rice field water, but products were not identified.

A 1976 survey by Crosby (*11*) revealed emerging evidence that photolysis was an important degradation process for certain classes of herbicides in several environmental compartments. Photosensitization was emerging as a significant factor in irradiation studies. It was speculated that in field studies, soil, water, and plant surfaces contained salts, solid oxides, fluorescent pigments, and other substances that could act to sensitize or desensitize photochemical processes.

Photolysis studies on soil surfaces proved extremely difficult to conduct and interpret. Soil studies would have been virtually impossible to conduct or interpret without the use of labeled compounds. There were suggestions that aqueous photolysis studies with and without photosensitizers would provide the same information as soil studies (*12*). EPA rejected this suggestion, arguing that the fate of photoproducts in water and soil would be different; that is, that photoproducts may hydrolyze in water but bind in soils or be metabolized in soil.

Important contributions to the early literature on soil photolysis studies were made by Koshy et al. (*13*) on procedures for photolysis on a solid surface and Niles and Zabik (*14*) on procedures for photolysis in aqueous solution, on soil, and on a thin film. An excellent review of the early literature on pesticide photolysis was prepared by Plimmer (*15*).

Metabolism. Probably the most difficult transitions from chemical to environmental technology were those studies with soils as the environmental matrix. This difficulty was particularly true of metabolism studies.

By 1970, most biochemists had some knowledge of xenobiotic metabolism in aqueous solutions. This knowledge was largely due to drug metabolism studies.

By comparison, understanding of microbial metabolism of pesticides was primitive, except for the phenoxy herbicides. Interpreting the results of metabolism studies in soils was complex. An important advance in soil metabolism studies was the development of soil biometer flasks by Bartha and Pramer (*16*). These simple two-chamber systems provided an effective means of trapping labeled CO_2 resulting from soil metabolism of the labeled pesticide. Biometer flasks required only 50 g of soil and were small enough to be manipulated for temperature or other environmental parameters. Another flask technique that allowed for simultaneous measurement of metabolism and volatility was developed by Kearney and Kontson (*17*). The atmosphere over the active soil containing labeled pesticide was drawn through a polyurethane foam plug that trapped volatile pesticides, while labeled metabolic CO_2 passed through the plug and was subsequently trapped in a basic solution. Other innovative techniques that were developed during this same time period made aerobic and anaerobic soil metabolism studies feasible from a chemical standpoint.

Current Guidelines

By 1982 the EPA had published a set of environmental guidelines entitled "Pesticide Assessment Guidelines, Subdivision N, Chemistry: Environmental Fate" (*12*). It was a far more extensive document than PR 70–15, and includes some 105 pages. Five major sections of the guidelines deal with environmental studies. These sections are further subdivided into several specific studies:

- Degradation studies: hydrolysis; photodegradation in water, soil, and air
- Metabolism studies: aerobic and anaerobic in soil and water
- Mobility studies: leaching and adsorption–desorption; laboratory and field volatility
- Dissipation studies: for field and aquatic uses (field studies), for forest uses, for combination products and tank mixes; and long-term soil studies
- Accumulation studies: confined studies on rotational crops, field studies on rotational crops, studies on irrigated crops, laboratory studies on fish, field studies on aquatic nontarget organisms

An extensive section is devoted to public comments on proposed regulations and then a detailed series of required studies. At the end of each section are referenced examples of published literature containing acceptable procedures. The references cited clearly show the impact that chemists in the Division of Pesticide Chemistry have had on pesticide regulation. The vast majority of this work was published in the early to mid-1970s, most of it appeared in the *Journal of Agricultural and Food Chemistry*, and the authors were members of the division. One of the division members' proudest accomplishments was to pave the way for sound and sane regulatory policy.

References

1. U.S. Department of Health, Education, and Welfare, "Report of the Secretary's Commission on Pesticides and Their Relationship to Environmental Health"; Parts I and II. U.S. Government Printing Office: Washington, DC, 1969.
2. U.S. Department of Agriculture, PR Notice 70–15, "Guidelines for Studies To Determine the Impact of Pesticides on the Environment"; U.S. Government Printing Office: Washington, DC, June 23, 1970.

3. James, A. T.; Martin, A. J. P. *Analyst* **1952,** *77,* 915.

4. *Analysis of Pesticide Residues*; Moye, H. A., Ed.; Chemical Analysis Series, Vol. 58; Wiley: New York, 1981; 467 pp.

5. Leng, M. L. In *Analysis of Pesticide Residues*; Moye, H. A., Ed.; Chemical Analysis Series, Vol. 58; Wiley: New York, 1981; Chapter 10, "Government Requirements for Pesticide Residue Analysis and Monitoring Studies", pp 395–448.

6. Krzeminski, S. F.; Brackett, C. K.; Fisher, J. D. *J. Agric. Food Chem.* **1975,** *23,* 1060–1068.

7. Gomaa, H. M.; Suffet, I. H.; Faust, S. D. *Residue Rev.* **1969,** *29,* 171–190.

8. Crosby, D. G.; Li, M-Y. In *Herbicide Photodecomposition in Degradation of Herbicides*; Kearney, P. C.; Kaufman, D. D., Eds.; Marcel Dekker: New York, 1969; Chapter 12.

9. "Symposium on the Use of Isotopes in Weed Research"; I.A.E.A.; Vienna, 1966; Slade, P. *Weed Res. 6* **1966,** *158,* 113.

10. Kuwahara, M.; Kato, N.; Munakata, K. *Agric. Biol. Chem.* **1965,** *29,* 880.

11. Crosby, D. G. In *Photochemistry in Herbicides: Chemistry, Degradation, and Mode of Action*; Kearney, P. C.; Kaufman, D. D., Eds.; Marcel Dekker: New York, 1976; Vol. 2, pp 836–890.

12. U.S. Environmental Protection Agency, Office of Pesticides and Toxic Substances; "Pesticide Assessment Guidelines, Subdivision N, Chemistry: Environmental Fate"; U.S. Government Printing Office: Washington, DC, 1982.

13. Koshy, K. T.; Friedman, A. R.; van der Slik; Graber, D. R. *J. Agric. Food Chem.* **1975,** *23,* 1084–1088.

14. Niles, G. P.; Zabik, M. J. *J. Agric. Food Chem.* **1975,** *23,* 410–415.

15. Plimmer, J. R. *Residue Rev.* **1969,** *33,* 47.

16. Bartha, R.; Pramer, D. *Soil Sci.* **1965,** *100,* 68–70.

17. Kearney, P. C.; Kontson, A. *J. Agric. Food Chem.* **1976,** *24,* 424–426.

Chapter 4

Consequences of Reregistration on Existing Pesticides

Marguerite L. Leng

In October 1988, the U.S. Congress approved amendments (*1*) to the Federal Insecticide, Fungicide, and Rodenticide Act (FIFRA) that mandated acceleration of the reregistration program authorized in 1972. The Environmental Protection Agency (EPA) was to reevaluate the scientific database underlying all active ingredients (AIs) contained in pesticide products registered before November 1, 1984. The concern was that old agrochemicals ("old" meaning those registered before 1984) may not have been tested adequately, and that they might be a hazard to humans or the environment. Any studies that did not meet current stringent standards for testing would have to be repeated, and the reregistration process must be completed within 9 years (by 1997). EPA began the reregistration program by dividing the process into five phases and setting deadlines for the completion of each phase.

Phase 1 of EPA's Reregistration Program

In Phase 1, EPA consolidated about 1300 active ingredients into about 600 "cases" that met defined criteria for the need to be reregistered. These cases were divided into four lists, which were each published (*2*) on schedule in 1989. Deadlines for completing each phase of the program were prescribed, as outlined in Table I.

**Table I. Current Deadlines for Completing the
U.S. Reregistration Program**

Phase	Responsibility	List A	List B	List C	List D
1	EPA	2/22/89	5/25/89	7/24/89	10/24/89
2	Registrant	?	8/25/89	10/24/89	1/24/90
3	Registrant	?	5/25/90	7/24/90	10/24/90
4	EPA	?	11/25/90	7/24/91	7/24/92
5	EPA within a span of 3 to 9 years after enactment of 1988 FIFRA amendments.				

List A included 194 cases covering all of the 350 active in-gredients for which EPA had issued Registration Standards be-tween 1980 and 1988. A Registration Standard is a published document that summarizes EPA's conclusions regarding the ad-equacy of the data supporting an active ingredient (or group of related compounds), identifies outstanding data gaps, and sets out the regulatory and label restrictions needed to protect health and the environment. One case or standard may include several active ingredients with similar chemical structures, such as the acid, salts, and esters of 2,4-D. List A includes most high-volume pesticides for agricultural use, food use, or home use, and ac-counts for at least 90% of the total volume of conventional pes-ticides used annually in the United States.

In mid-1989, registrants of products containing chemicals on List A were surveyed by the National Agricultural Chemicals Association (NACA) and by Interregional Project No. 4 (IR-4, which represents all 50 states and territories) to estimate how much support could be expected for all uses of the 194 cases for which EPA had prepared and issued Registration Standards. These surveys (3, 4) revealed that all current food or feed uses would be canceled for at least 20 cases, one or more food or feed uses would be dropped for 39 cases, and all current food or feed uses would be supported for only 53 cases.

Meanwhile, EPA conducted an inventory of the data sup-porting chemicals on List A, and compared the data require-ments levied in each Registration Standard with data requirements imposed under current testing guidelines. EPA also evaluated the acceptability of new studies submitted in re-sponse to any "data call-in" under each Registration Standard

(5). As outlined in Table I, no deadlines were set for completion of each phase of the reregistration program for List A pesticides, although they represent the greatest volume of use, particularly on food crops. Table II outlines the status of List A pesticides in March 1990, and indicates that sufficient data were on hand for reregistration of only 11 of the 194 cases. However, only two of these were actually reregistered by the end of 1990: *Heliothis zea* with only one registered product and one registrant, and fosetyl Al (Aliette) with two registered products and one registrant (6).

List B included 149 cases representing 229 active ingredients contained in 14,000 registered formulations. These cover pesticides not on List A with food or feed uses, others used on crops, those with potential toxicological concerns, and those with outstanding data gaps. Some of these cases represented very old products no longer being used, but whose registrations were still on file at EPA in 1989 when the lists were compiled.

List C included 150 cases representing 288 active ingredients contained in 11,000 registered products. Many are antimicrobials or disinfectants that EPA regulates as pesticides, and some are solvents or adjuvants that are claimed to have some pesticidal properties. In general, fewer studies are required to support registered uses for such products compared to requirements for pesticides used on food or feed crops or in extensive areas

Table II. March 1990 Status of List A Pesticides
(194 Cases, 350 Active Ingredients)

Cases	Results of EPA's Inventory of Supporting Data
28	All products and uses have been cancelled or suspended
11	All required studies are on hand; can be reregistered in FY 90
53	Awaiting submission of data, but may be reregistered in FY 91–94
51	Candidates for further data call-in; notices are to be issued in FY 90
30	Need administrative tracking of overdue studies; to be completed by 3/90
21	Need in-depth review of submitted studies; data call-in by 5/90

that are not used for growing crops (e.g., noncropland like rec-
reational areas; rangeland; and rights of way along roadsides,
pipelines, and utility lines).

List D included the remaining 118 cases, representing 286
active ingredients. It was organized in three parts: conventional
chemicals, microbials, and biochemicals, few of which are used
in food or feed crops.

Phase 2 of EPA's Reregistration Program

In Phase 2, registrants had to submit responses listing all available studies judged to be acceptable to EPA, and commit to conduct all additional studies needed to meet EPA's current
requirements for registration of a new product. The burden was
considerably less for registrants of formulated products than for
basic manufacturers of active ingredients who have the responsibility for providing all of the studies needed for a complete
generic database supporting each active ingredient.

This burden was magnified by EPA's new requirement for
separate studies to support each minor variation in the active
ingredient, such as an acid and each of its salts and esters that
are contained in individual registered products. For example,
EPA has requested a number of repetitive studies to support the
33 salts and esters of 2,4-D that had been registered during more
than four decades since the herbicidal property of the acid was
noted in 1945. It appears that EPA wants to ascertain whether
any minor differences can be detected in data from similar studies on the toxicity, metabolism, residues, or rate of degradation
of each individual derivative compared to data from studies on
the acid or of a representative ester that is converted rapidly to
the acid. As a result, an Industry Task Force composed of basic
manufacturers of 2,4-D agreed to share the cost of testing only
11 of the derivatives. Consequently, EPA sent "Notices of Intent
To Suspend" to registrants of at least 100 products containing
any of the other 22 derivatives of 2,4-D. Most of the suspended
products had been registered for specialty uses and are now lost
to the agricultural industry.

As indicated in Table I, registrants had only 90 days after
publication of each list to submit a response to EPA in support

of each registered product containing an active ingredient on Lists B, C, or D. In mid-1989, EPA issued guidelines (7) for preparation of Phase 2 responses, including lists of all studies deemed necessary for reregistration of any pesticides on List B. Registrants had to provide information on each of the following for each registration:

1. specifying "yes" or "no" on seeking reregistration for the product

2. identifying all studies required to support that registration

3. evaluating existing studies as to acceptability according to current testing standards

4. identifying whether each study judged to be adequate had been submitted and was on file at EPA, and providing its unique eight-digit EPA Master Record Identification Number (MRID#)

5. committing to conduct new studies that were needed because

   • no previous study exists (true data gap)

   • a previous study was conducted before 1970

   • raw data for a previous study are not available

   • an existing study does not meet current EPA guidelines, even though it was conducted after 1970 and raw data are available

6. providing payment of the first portion of the reregistration fee (1)

EPA had estimated that the average burden for completing a Phase 2 response was only 42 hours. However, it took considerably longer for affected basic manufacturers to locate and evaluate the adequacy of many studies that had been submitted previously in support of registration for all uses of each product containing pesticides on Lists B, C, and D, and to decide whether it was worth the effort to maintain each registration. The estimated costs were enormous for repeating all studies judged to be inadequate by current testing standards, and also for conducting many new studies to fill new requirements imposed by EPA in recent years. Furthermore, most older pesticides had lost their status as proprietary products, so consortiums had to be formed of registrants or manufacturers willing to share the high costs of generating new data on these generic chemicals. Added to the difficulty of such decisions was the uncertainty that EPA would even accept many of the old studies that the registrants deemed to be adequate, and that no further studies would be added to the already overwhelming list of requirements to support registration of an active ingredient and its minor variations.

In mid-1989, EPA estimated that List B represented at least 8000 products manufactured by 3500 companies, and that about 40,000 to 50,000 new studies would have to be reviewed for 229 active ingredients included in the 149 cases. Detailed guidelines for Phase 2 responses that were issued to registrants by EPA on May 19 defined as many as 134 individual studies that a basic manufacturer would have to submit to support reregistration of an active ingredient used on at least one food crop. The number and kind of studies generally needed to fulfill all of the requirements under each of the subdivisions in the guidelines for testing pesticides are as follows:

- 26 on product chemistry (manufacturing process, analysis, and properties)
- 27 on toxicology in mammals (including chronic lifetime feeding studies at maximum tolerated doses to test the potential for causing cancer, multigeneration reproduction studies, and teratology studies on pregnant animals to test the potential for causing birth defects)
- 27 on toxicity to wildlife (birds, fish, and aquatic organisms)
- 15 on toxicity to nontarget plants and nontarget insects
- 24 on environmental fate of active ingredient and all its degradation products
- 2 on extent of drift of applications from the target area

- 13 on nature of the residues in crops and analytical methods to measure them
- 5 on residues resulting from treatment of each crop or each test site

Some of these requirements are new or have been modified to add new specifications. It is a very great commitment for any manufacturer to agree in advance to provide all of these studies, regardless of the sales volume for any active ingredient. It is overwhelming for a small company with only old products that were first registered many years ago.

In May 1990, EPA reported (*8*) that no support had been provided for many of the active ingredients on Lists A, B, and C. Those unsupported on List B included 53 active ingredients contained in 305 products with Federal registrations under Section 3 or for Special Local Needs (SLN) under Section 24(c) of FIFRA, and 29 active ingredients had no active registrations. Many additional active ingredients were dropped by manufacturers during preparation of Phase 3 responses, and others will probably be suspended by EPA following preliminary review of those responses during Phase 4 of the reregistration program.

Phase 3 of EPA's Reregistration Program

According to very detailed guidelines for Phase 3 responses issued by EPA in December 1989 (*9*), registrants must reformat every pre-1982 study that they want EPA to consider in support of a pesticide reregistration. Each study must be bound separately and must use EPA's very strict arrangement for an overall sequential page-numbering system beginning with a new title page; a nonconfidentiality statement signed by the sponsor and submitter; a Good Laboratory Practice (GLP) statement signed by the study director, the sponsor, and the submitter; a complete table of contents listing all tables, figures, and appendixes; and a detailed summary. The summary must address each guideline requirement for that study, and provide explanations why any of the many "acceptance criteria" are not fulfilled. A special "flagging" page is also inserted after the GLP statement for all residue and toxicology studies (except acute and short-term subacute studies), so EPA can spot immediately whether any adverse effects were noted in any of these studies.

Each Phase 2 and Phase 3 response also must be accompanied by a large fee, as defined in the 1988 amendments to FIFRA (1), to defray any costs EPA may encounter in having the studies reviewed, either by EPA personnel or by outside contractors. The total reregistration fee is $150,000 for a major pesticide with food or feed crop uses, and $75,000 to $100,000 for an active ingredient with nonfood–feed uses. In cases where a pesticide is produced by more than one manufacturer, this fee is apportioned among them according to their estimated U.S. market share.

Phases 4 and 5 of EPA's Reregistration Program

EPA's task in Phase 4 is to give a quick review to all reports submitted under Phases 2 and 3 to see that all required studies are accounted for and to identify any obvious data gaps (1). The term "data gap" often refers to minor deficiencies in studies that can be upgraded to meet ever-changing requirements according to the latest guidelines. Unfortunately, no one can anticipate what tomorrow's guidelines will require; yet registrants have to commit now to conduct whatever studies EPA may eventually say are needed, including entirely repeating many studies. On January 10, 1991, EPA issued an update indicating that reviews of Phase 2 and Phase 3 responses for List B pesticides were 6 months behind schedule, but that they expected to issue detailed data call-ins (DCIs) for all remaining pesticides by June 1991 (6). The status on March 8, 1991, is outlined in Table III

Table III. List A, B, C, and D Pesticides
Supported over Time

List	December 1988		March 8, 1991		Already Lost	
	AIs	Cases	AIs	Cases	AIs	Cases
A	350	194	293	159	57	35
B	229	149	141	109	88	40
C	288	150	124	83	164	67
D	286	118	113	66	173	52
Total	1153	611	671	417	482	194

adapted from EPA's information pamphlet entitled "Pesticide Reregistration" (*10*).

EPA has also set very tight deadlines for completing and submitting all additional studies. These deadlines may be hard to meet because of limitations in the number of available testing facilities that are capable of performing studies according to strict new GLP guidelines for all chemistry and field studies (*11*), as well as earlier GLP standards for toxicology studies (*12*).

In Phase 5, within a span of 3 to 9 years after enactment of the 1988 FIFRA amendments, EPA has the enormous task to conduct a thorough and comprehensive evaluation of all studies submitted in support of registration and reregistration of all remaining active ingredients. Product specific data for individual formulations would be due 8 months later, and must be reviewed within 90 days. The ultimate goal of reregistration of a product should be accomplished by 6 months after that, presumably by the 1997 deadline set by Congress (*1*).

Fees for Maintaining Registrations

The 1988 amendments to FIFRA did away with the fees EPA had proposed to collect for each registration action requiring review of data. Instead, EPA was authorized by Congress to require payment of annual maintenance fees for each registration (*1*). Each registrant was assessed $425 per product for 1 to 50 products, or $100 per product for 51 to 200 products per year. The maximum annual fee was $20,000 for a company with 50 or fewer products, and $35,000 for one with more than 50 products. Included were many Section 24(c) state registrations for Special Local Need (SLN) labels for minor uses requested by growers of specialty crops in those states.

These annual fees were intended to support the added work needed for EPA to review, in a timely fashion, all the studies submitted during the reregistration process so as to accomplish reregistration by the 1997 deadline imposed by Congress. However, the income derived from these fees in 1989 was much lower than expected because many registrants were not willing to pay a separate fee for each variation of a product label. As a result, EPA canceled more than 20,000 registrations, and growers lost many products they had counted on for control of pests in minor crops (*13*).

To make up for this income deficit, EPA tripled the rates for annual maintenance fees in 1990. Legislation has also been proposed that would eliminate the upper limit (or "cap") for registrants with multiple products (*14*). This step can only lead to the loss of more products and greater impact on growers of minor crops. Registrants cannot justify paying the continually escalating costs of maintaining registration for all of their products, particularly for older products that are no longer covered by patents but are sold in competition with newer patented products (*15*). Consumers are the eventual losers when fewer and more costly products are available to the growers, particularly for minor crops, which include most fruits and vegetables.

Consequences of the Current Reregistration Process

The American farmer has already lost the use of many familiar products (*8, 13*) and faces the loss of many more, particularly for minor uses that registrants cannot afford to support. Also of concern is the anticipated delay in getting new products tested and on the market, because limited laboratory facilities are already booked to full capacity with repeating old studies or conducting new studies to meet ever-expanding requirements for reregistration of old products. Additional aspects of EPA's reregistration process of concern to the agrochemicals industry are the following:

1. Many studies will have started or been completed before GLP guidelines were published in final form on August 17, 1989 (*12*).

2. Evaluation of studies will be on a guideline-by-guideline basis, by contractors, using a checklist approach for comparing data against very detailed acceptance criteria outlined in the voluminous Phase 3 Guidance Package issued by EPA in December 1989 (*9*).

3. Registrants will be given deadlines of either 1, 2, 3, or 4 years for submitting additional studies, depending on the time EPA estimates is needed to complete individual studies (7).

4. Notices of suspension or cancellation of product registrations will be published by EPA in the *Federal Register* if individual studies have not been submitted by specified deadlines, and extensions will be granted only if the registrant can show that delays were unavoidable.

5. Reregistration is a moving target without a defined end (16).

During 1989, certain companies objected when EPA issued notices of intent to suspend their products because of failure to meet deadlines for submitting required studies. Actually, the studies had been submitted and "logged in", but EPA's Registration Division had not informed their Enforcement Division that the studies had already been accepted for review. Subsequently, EPA instituted a computer tracking system to avoid such problems and to alleviate the concern of those registrants who strive to comply with unreasonable deadlines.

The "moving target" issue is of great concern in that a diligent registrant may have completed all the studies to fill data gaps identified by EPA during a first-round review, but may later receive an even larger list of gaps identified during a second-round review. In an attempt to resolve this impasse, NACA requested EPA to provide an official definition of what constitutes reregistration. NACA proposed that (16)

It should be viewed as a process with a beginning, a middle, and an end, with a fixed set of criteria for each pesticide so that one can tell how to reach the end of the process. Reregistration of a pesticide should mean that the Agency has taken a modern, in-depth look at the pesticide and concluded that its database is reasonably near completion and that the pesticide appears to satisfy the FIFRA criteria.

EPA held a workshop on September 24–25, 1990, to resolve this problem, but to date (March 14, 1991), no definition of reregistration has been issued.

Costs of Developing a Pesticide for Use in the United States

The cumulative cash flow for a hypothetical new pesticide is outlined in Figure 1. The cost is low during initial test stages after discovery, but escalates rapidly as data needed for registration purposes are being developed. Assume that a patent is obtained (A) in year 3, that registration is granted (B) in year 7, and that a production plant has been built (C) by year 10. A total of $70 million may have been expended before any upturn is seen in the cash flow line. If the new product succeeds in a competitive market and sales are good, the positive side of the ledger may be reached (D) by the 15th year after discovery of this new pesticide. FIFRA as amended in 1988 requires reregistration 9 years after the first registration (E). However, on the basis of present value at an 8% discount rate, the first real dollar of profit is not achieved until the 20th year after discovery (F).

Unfortunately, U.S. patents protect the rights of owners for only 17 years after being granted. Consequently, other manufacturers, both domestic and foreign, can now obtain registrations for this proven product and enter the market without the potential risk of failure for a new product. The sales price will drop because U.S. regulations permit the use of old data for these "me-too" products* without compensation to the submitter of the data. The original registrant now has a lower rate of return on his investment, but is still faced with added costs for reregistration of the product. This burden may be the last straw needed to convince the original registrant to drop the product. Then newcomers in the market must either generate all the data needed for reregistration or also drop out.

Another disincentive for American registrants is that data requirements are much fewer in certain other countries, and often there is greater flexibility as to what constitutes an acceptable study for registration purposes. Thus, the potential for success is much greater in foreign countries, and an adequate

*A me-too product is one that is represented as equivalent or substantially similar to a previously registered product.

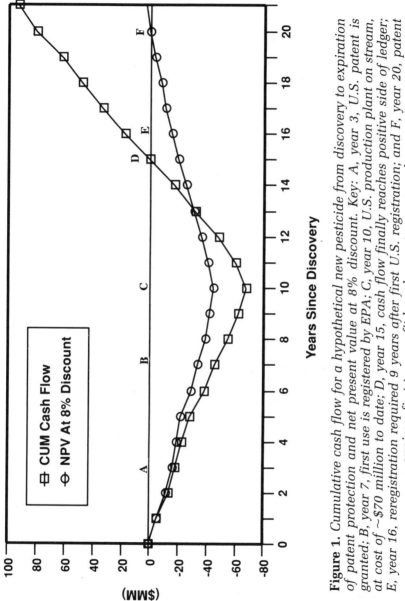

Figure 1. *Cumulative cash flow for a hypothetical new pesticide from discovery to expiration of patent protection and net present value at 8% discount. Key: A, year 3, U.S. patent is granted; B, year 7, first use is registered by EPA; C, year 10, U.S. production plant on stream, at cost of ~$70 million to date; D, year 15, cash flow finally reaches positive side of ledger; E, year 16, reregistration required 9 years after first U.S. registration; and F, year 20, patent expires; first true profit based on net present value.*

margin of profit can be accomplished sooner than in the United States.

Ranking of Agrochemical Companies in Worldwide Industry

History has shown that the biggest agrochemical companies have headquarters in Europe, and that many smaller companies have been lost as a result of acquisition or merger with larger companies. In view of escalating registration costs in the United States and the need to remain competitive in the world market, several smaller U.S. agrochemical companies are looking for partners or expect to be sold by their parent chemical or pharmaceutical companies.

Table IV lists the top 20 basic pesticide manufacturers in order of total dollar sales worldwide in 1989 and their relative rank in recent years. Footnotes indicate acquisitions and mergers that had an impact on sales and relative rankings. The top U.S.-based company is Du Pont, which advanced from ninth to fifth place by acquiring Shell's U.S. agrochemicals business in 1986. As a result, Shell dropped from sixth to tenth place worldwide in recent years. In 1989, the Global Agricultural Chemicals Department of The Dow Chemical Company entered into a joint venture with the agricultural chemicals portion of Eli Lilly's Elanco Division. This new company, named DowElanco, was ranked seventh in 1989, an advance from 10th and 15th, respectively, in 1987. Another recent joint venture between Sumitomo (Japan) and Chevron (USA) is known as Valent for products sold in the United States, while Chevron acquired products formerly registered by PPG (ranked 40th worldwide in 1987). Pennwalt had advanced from more than 40th in 1986 to 20th in 1989, but merged with Atochem and M&T in 1990. Also in 1990, Ciba–Geigy acquired Dr Maag (17th in 1989) and is still in first place worldwide. Also of concern is that 12 Japanese companies were ranked in the top 40 worldwide in 1988, and have many new products that do not have the burden of requiring new data for reregistration. In addition, the first acquisition move by a Japanese chemicals firm in the U.S. pesticides market

Table IV. Ranking of Major Agrochemical Companies Based on Total U.S. Dollar Sales Worldwide

Company (Headquarters)	1989	1988	1987	1986
Ciba Geigy (Switzerland)	1	1	1	2
Bayer (Germany) and Mobay (USA)	2	3	2	1
ICI (United Kingdom), Stauffer (USA)[a]	3	2	3	5
Rhone Poulenc (France), Union Carbide (USA)[b]	4	4	4	3
Du Pont (USA), Shell (USA)[c]	5	5	5	9
Monsanto (USA)	6	6	6	4
DowElanco (from Dow, USA and Elanco USA)[d]	7	10	10	10
BASF (Germany)	8	7	8	7
Hoechst (Germany)	9	9	9	8
Shell (Netherlands, without Shell USA)[c]	10	8	7	6
Schering (Germany), FBC (UK), NorAm (USA)[e]	11	11	11	11
American Cyanamid (USA)	12	12	13	12
Sandoz (Switzerland), Zoecon (USA), Velsicol (USA)[f]	13	13	12	13
FMC (USA)	14	16	17	18
Rohm & Haas (USA)	15	19	20	17
Uniroyal (USA)	16	23	26	25
Dr Maag, Roche (Switzerland)[g]	17	24	25	24
Makhteshim (Israel)	18	27	29	27
Fermenta (Sweden)	19	28	28	28
Pennwalt (USA)[h]	20	29	33	>40

[a]Imperial Chemicals Company acquired Stauffer's agricultural chemicals in 1986.
[b]Rhone Poulenc acquired Union Carbide's agricultural chemicals in 1985.
[c]Du Pont acquired Shell's U.S. agricultural chemicals in 1986.
[d]Dow's agricultural chemicals and Eli Lilly's agricultural chemicals merged in 1989.
[e]Schering acquired FBC and NorAm prior to 1986.
[f]Sandoz acquired Zoecon in 1983, and acquired Velsicol in 1986.
[g]Ciba Geigy (Switzerland) acquired Dr Maag (Switzerland) in 1990.
[h]Pennwalt (USA) and M&T (USA) joined Atochem as Atochem North America in 1990.

was the late 1990 purchase by Ishihara Sangyo Kaisha of Osaka. Sweden's Fermenta sold Ishihara Diamond Shamrock's former facility in Ohio, known as SDS Enterprises, and its fungicide subsidiary, Fermenta ASC, as well as Ricerca, a contract R&D operation. The new company is ISK Biotech Corporation (*17*).

Conclusion

A major objective of the 1988 FIFRA amendments was to reestablish a credible U.S. reregistration program by expediting the reevaluation, retesting, and determination of eligibility of all currently marketed products. Many products have now been lost for use by the American farmer, and many more will likely be lost before reregistration has been accomplished. Third-party registrations of end-products for minor uses cannot be accomplished unless a basic manufacturer has provided all the generic data required to support reregistration of the active ingredient. Furthermore, the high costs for registering new products or reregistering old products in the United States are forcing smaller agrochemicals companies to either drop out of the business or to join forces with other U.S. or foreign companies. This situation may also lead U.S.-based companies to move overseas, with resultant loss of jobs and revenue for many Americans.

Acknowledgments

The author acknowledges the following persons for providing key information used in updating this chapter, originally presented in Miami during the ACS Fall 1989 National Meeting:

1. Jay S. Ellenberger, Chief of Generic Chemical Support Branch, Special Review and Reregistration Division, Office of Pesticides and Toxic Substances, EPA, Washington, DC.

2. Carol Stangel, Special Review and Reregistration Division, Office of Pesticide Programs, EPA, Washington, DC.

3. Ray S. McAllister, Director of Regulatory Affairs, National Agricultural Chemicals Association, Washington, DC.

4. Mark A. Wegenka, Business Analyst, DowElanco, Indianapolis, IN.

References

1. "Federal Insecticide, Fungicide, and Rodenticide Act", enacted by Public Law 92–516, *U.S. Code* Title 7, Pt. 136 et seq., October 21, 1972, as amended through October 25, 1988, by Public Law 100–532.

2. "Pesticides Required To Be Reregistered", List A, *Federal Register* 54 (February 22, 1989); List B, *Federal Register* 54 (May 25, 1989); List C, *Federal Register* 54 (July 24, 1989); List D, *Federal Register* 54 (October 24, 1989). Also, *Farm Chemicals Handbook '91;* Meister Publishing: Willoughby, OH, 1991; pp D10–D13.

3. "NACA Survey of Food–Feed Crop Uses Dropped for List A Pesticides" and "NACA Survey of Food–Feed Crop Uses Dropped for List B Pesticides", National Agricultural Chemicals Association: Washington, DC, October 17, 1989.

4. *IR-4 Pesticide Reregistration Alert;* Interregional Research Project No. 4, New Jersey Agricultural Experiment Station: New Brunswick, NJ, issued periodically.

5. "Reregistration Plan for List A Announced; Activities Scheduled for 194 Chemical Cases", *Chemical Regulation Reporter, Current Report* March 23, 1990, p 1607. Details of the status are provided in a letter dated April 3, 1990, to Ray McAllister, National Agricultural Chemicals Association, from Margaret Stasikowski, Special Review and Reregistration Division, EPA, Washington, DC.

6. *Pesticide Reregistration Handbook: How To Respond to Reregistration Eligibility Document (RED);* Special Review and Reregistration Division, EPA, Washington, DC, January 1991. Details of the status of the reregistration program are provided in a letter dated January 10, 1991, to interested parties from Allan S. Abramson, Special Review and Reregistration Division, EPA, Washington, DC.

7. "Guidelines for Preparation of Phase 2 Responses", issued May 19, 1989, by Douglas D. Campt, Office of Pesticide Programs, EPA, to registrants of List B pesticides, and amended on August 10 (just before the August 24, 1989, deadline) to provide new response forms and to specify new completion dates for additional studies.

8. Letter dated May 14, 1990, to Marguerite Leng from Jay S. Ellenberger, Special Review and Reregistration Division, EPA, Washington, DC.

9. "FIFRA Accelerated Reregistration—Phase 3 Worksheet and Guidance", issued April 12, 1990, by Jay S. Ellenberger, Special Review and Reregistration Division, EPA, Washington, DC.

10. "Pesticide Reregistration", an information pamphlet issued by the EPA, in press, April 1991.
11. FIFRA Good Laboratory Practice Standards, Final Rule, EPA, *Federal Register* 54 (August 17, 1989) pp 34053 ff.
12. FIFRA Good Laboratory Practice Standards, *Code of Federal Regulations* Title 40, Section 160. Also *Federal Register* 48 (November 29, 1983) pp 53946 ff.
13. "List B Active Ingredients Unsupported by Registrants", *Pesticide and Toxic Chemical News* June 13, 1990, pp 43–44. "List C Active Ingredients for Which Generic Data Are Not Being Supported", EPA: Washington, DC, May 24, 1990.
14. "Remove Maintenance Fee Caps, EPA Suggests to Rep. de la Garza", *Pesticide and Toxic Chemical News* March 28, 1990, p 36. Details of the proposed increase in maintenance fees were provided in a letter dated February 27, 1990, from Linda J. Fisher, EPA, to the Honorable E. de la Garza, House of Representatives, Washington, DC.
15. "Industry: FIFRA Reform Needed To Avoid High Reregistration Fees", *Inside EPA Weekly Newsletter* January 5, 1990, p 10. The fees were originally set under FIFRA as amended in 1988.[1]
16. Letter dated June 30, 1989, from Ray S. McAllister, National Agricultural Chemicals Association (NACA), to Edwin F. Tinsworth, EPA, proposing a definition for "Reregistration of Pesticides under FIFRA".
17. *Chemical Week* October 24, 1990, p 15.

Chapter 5

A Short History of Pesticide Reregistration

Edward C. Gray

Much of the public's concern about pesticides is based on the awareness that some pesticides now in use were first registered long ago and never have had a modern evaluation. The Federal law that governs the sale and use of pesticides, the Federal Insecticide, Fungicide, and Rodenticide Act (FIFRA) (1), requires that these older pesticides be reevaluated through a process called "reregistration". In this chapter, I will describe the concept of reregistration, how and why it arose, and how it has changed over the years.

Reregistration as we think of it today has three main components:

1. updating the database on a registered pesticide by determining the current data requirements and ensuring that data are available to fill those requirements

2. reevaluating the data against the current registration criteria to decide if changes in the pesticide's registration are needed

3. requiring modifications in the use instructions and precautions that appear on the pesticide's labeling, or canceling the pesticide's registration

2085-9/91/0045$06.00/0 © 1991 American Chemical Society

The FIFRA Registration Scheme

The need for a reregistration process arises from the nature of the FIFRA registration scheme, which dates from FIFRA's enactment in 1947 (2). Under FIFRA, a pesticide product may not be lawfully distributed or sold in this country unless the U.S. Environmental Protection Agency (EPA) has issued a license called a "registration" for it (3). Moreover, since 1972 FIFRA has made it unlawful for any person to use a pesticide product in a manner inconsistent with the labeling that is approved as a part of the product's registration. A great deal of importance thus attaches to the initial registration of a pesticide, and the statute reflects this by specifying detailed procedures and criteria governing the issuance of initial registrations.

An application for a pesticide registration must contain the product's proposed labeling, which tells the user how, where, and when the product may be used, and what precautions must be observed. The applicant for registration must also make available to EPA a variety of data on the composition, toxicity, and environmental behavior of the pesticide, and—if it is intended for use in food production—on the nature and level of pesticide residues on food items that result from its use. EPA evaluates the data and the proposed labeling and decides whether the product meets the FIFRA risk–benefit criterion for registration. EPA often refuses to grant a proposed registration unless changes are made in the proposed labeling to reduce the product's risk potential or clarify the use instructions.

This premarket review system has two obvious but important limitations:

1. It applies only to products initially entering the market.

2. It necessarily measures proposed products against the statutory provisions, data requirements, and safety criteria that are in place at the time the application is being considered.

Congress has made several basic changes in the scope and coverage of FIFRA since its original enactment (4). Moreover, the prescribed data requirements have grown considerably, as new technologies are developed in areas ranging from radiochemistry and chromatography to behavioral neurotoxicity. Finally, regulatory concerns have broadened and changed, and the registration criteria have been interpreted and refined accordingly (5). These kinds of changes logically should be imposed on already-registered products at the same time they take effect with regard to new pesticides (6). But a premarket review system by itself provides no means to ensure that the registrations of older products will be modified as necessary to reflect EPA's new data requirements or revised criteria.

Accordingly, a method for requiring changes to already-issued registrations is a necessary part of any licensing system like FIFRA's. It should not be surprising that this need occurred to Congress when FIFRA was first being considered. The 1947 law provided for a process designed to accomplish a type of "reregistration". The law authorized the Secretary of Agriculture to cancel any registration on the fifth anniversary of its issuance unless the registrant, in accordance with regulations issued by the U.S. Department of Agriculture (USDA), requested that the registration be continued. The House Committee on Agriculture, in its report (7) on the bill that became the 1947 law, said, "This provision will have the effect of keeping registrations up to date."

FIFRA did not prescribe any particular process for USDA to use in deciding whether a request for continuation of a registration had been made properly or whether cancellation for failure to do so was justified (8).

USDA never took to heart the Congressional drafters' notion that FIFRA pesticide registrations were time-limited licenses that should be reexamined periodically and renewed if that was justified. The 5-year renewal requirement was used at USDA primarily as a means of removing from the registration rolls those products that registrants no longer cared about. Occasionally the Department sought to use the requirement to remove products from the market on health grounds (9). But it does not appear that USDA ever used the renewal provision to modernize the databases for active registrations. During the 1960s, as more data requirements were imposed and more products registered, USDA fell farther and farther behind in dealing with renewal, and all but abandoned its use in the later part of the decade.

In 1964 FIFRA was amended (*10*) to add a process for canceling registrations that were thought to pose unacceptable risks. This cancellation process was independent of the 5-year renewal requirement, which was not changed by the 1964 amendments. The cancellation process Congress adopted was very protective of registrants' procedural rights; it allowed registrants to demand both a referral to an advisory committee and a formal, trial-like, adjudicatory hearing. USDA promptly issued regulations that extended the formal hearing requirement to disputes about 5-year renewals, thus abandoning the idea that a registration is a temporary right to sell a pesticide, a right that could be renewed periodically but only if it was periodically reassessed and found to satisfy current requirements. In its place came the idea of registration as a sort of permanent, propertylike right that can be taken away only by means of a courtroom proceeding, an idea that we are still familiar with under today's law and practice. The greater permanence of this kind of registration makes the pesticide industry more willing to invest in research and development. This increased protection comes at some cost, however. Critics argue that the cost and time of hearings make it difficult for EPA to adequately regulate existing pesticides when new evidence of risk arises.

During the late 1960s, USDA came under increasing pressure to regulate registered pesticides more efficiently. One approach USDA used was issuing interpretative regulations stating a position on whether a particular kind of pesticide should be marketed and thereafter instituting cancellation actions to carry out the policy set forth in the interpretation. Such regulations often concerned acute toxicity problems (*11*).

USDA also was increasingly urged to review the alleged chronic health and environmental effects posed by persistent "old" pesticides. Lawsuits by environmental groups seeking cancellation of DDT and 2,4,5-T led USDA to conclude that some pesticides were too environmentally dangerous to be used without restrictions, but that trained users who would follow directions could avoid many of the problems caused by the pesticides. USDA also noted that to make such an approach defensible and workable, FIFRA would have to be amended. The Administration developed a legislative proposal for classifying pesticides into categories of general use, restricted use, or use only by permit, and for making these classifications (and other label directions) enforceable at the retailer–user level. This legislation was introduced in early 1971, just after EPA had been formed and had taken over the pesticide program. The result was the 1972 FIFRA amendments (*12*) and more changes in the idea of reregistration.

The Reregistration Program

According to conventional wisdom, our modern reregistration program was mandated by the 1972 amendments. That is true in one sense but not in another. Certainly Congress did decide in 1972 to regulate the use of pesticides, not just the sale, and to require that all pesticides should be classified as either for restricted use or for general use and be labeled as such. In most cases, restricted-use pesticides were to be usable only by trained, certified applicators. This meant that each one of the 30,000 or 40,000 products had to be subjected to a review to decide the product's proper classification. It also meant that each product had to be relabeled so that anyone could tell from the label whether the product could legally be used by an uncertified applicator. The law said that by 1974 EPA must establish regulations governing registration and classification, and reregister all the products between 1974 and 1976.

In 1972 both Congress and EPA innocently assumed that a swift review of a product's registration folder would reveal any problems associated with the product, and that the classification of the product ordinarily would be obvious. This assumption proved to be incorrect. As EPA developed its registration and classification regulations, which included a specification of the kinds of data required for various kinds of registrations, it became obvious that the process of amassing and evaluating a modern database would take a great deal longer than the time Congress originally had provided.

These registration regulations, issued in mid-1975 (*13*), described a very ambitious regulatory program. For each pesticide, all currently available data would be assessed against specific criteria with respect to whether they should be canceled and, if not, how they should be classified. For products with missing

long-term studies, EPA planned to issue reregistrations that
would be valid only for a reasonable period of time not to exceed 5 years. Each registrant would have to apply for the reregistration of each of its products, and any deficient application
would be denied. All this would be accomplished in a year and
a half. These regulations also attempted to resuscitate the long-
dormant 5-year renewal requirement; EPA said that any product
would be canceled on its anniversary unless EPA had concluded
that the registrant had satisfied all requirements of the Act and
the new regulations, including having a complete set of supporting data.

Within a year it had become obvious that this scenario was
totally unrealistic. The current database was found to be much
less useful than had been assumed, and the resources for conducting all the promised reviews and evaluations simply were
not available.

From 1976 until 1988, EPA obtained a great deal of new
information and used it to persuade registrants to amend their
registrations in many respects. In terms of actually completing
the process, however, during that period the history of reregistration largely consisted of extensions of projected dates for the
initial reassessment of all registered pesticides and decreases in
the complexity and scope of the reregistration program, to match
the resources available. EPA obtained amendments to FIFRA in
1978 that removed the deadlines for reregistration, severed reregistration from classification, emphasized evaluation of active
ingredients rather than individual products, and gave the EPA
clear authority to require submission of additional data regarding existing products.

In 1979 EPA issued an advance notice of proposed rulemaking (14) describing how it would issue "registration standards"
under the 1978 amendments and how they would be used. Compared with what was envisioned by the 1975 regulation, the
1979 description of reregistration is modest in scope. Even so,
many of the features of reregistration as described in the 1979
document have never materialized or were quickly abandoned
in order to save resources. As seen by EPA in 1979, a registration
standard was to be a general-purpose document that would ex-
plain what EPA thought about products containing a particular
active ingredient. The standard would state which existing

products should be canceled or have their labeling modified, and would be used to determine which new products could be registered. It would set forth the allowable uses and labeling for any product containing a particular active ingredient, and would be updated from time to time.

EPA did not follow this approach, either. Standards are not much used in the day-to-day registration process, and have not become the principal tool for determining registrability. For many active ingredients, registration standards have consisted primarily of requirements for the submission of further data. There has been little use so far of registration standards as a basis for cancellation actions for products whose labeling is at odds with what the standard prescribes. Most important, EPA has hardly begun to evaluate end-use products in the registration standards process; the focus has been almost exclusively on the active ingredients. The latest regulations on the subject of reregistration were proposed in 1984 but not issued in final form until 1988 (*15*).

Congressional and public concern with the resource-constricted pace of reregistration led to further legislative activity on the subject from 1985 to 1988, culminating with the enactment of amendments to FIFRA in October 1988 (*15*). These amendments contain some 13 pages of provisions on how to conduct reregistration. They do not, however, change the basic elements of reregistration—updating the database, subjecting it to a modern review, and implementing needed changes in the registrations. In the main, the amendments simply set a schedule for the completion of the data gathering and data review steps of the process, and provide resources—fees to be paid by registrants, and work to be performed by them—to allow EPA to finish the reevaluation process somewhat sooner. And the law stops short of describing what the EPA is to do once it concludes that a product should not be reregistered; it merely states that EPA shall take appropriate action under other parts of FIFRA. The amendments do require EPA to conduct a reregistration assessment of each end-use product, something that the EPA has deferred repeatedly to save resources. The initial impact of the 1988 amendments has been drastic, however, as registrants voluntarily dropped tens of thousands of unpromising registrations to avoid fees and data development costs.

The Future

In 1989 Congress again began to consider amendments to FIFRA. EPA has argued that Congress should simplify the provisions for canceling registrations. It has proposed that a process resembling informal rulemaking should replace the current process— a combination of an informal (and often very lengthy) "special review" investigation followed by a formal adjudicatory hearing process. EPA has claimed that a shorter cancellation process is needed to deal credibly with the prediction that a number of pesticides will be found to be unreasonably hazardous as a result of the reregistration process. Opponents argue that formal adjudication should be retained because it ensures that EPA will attempt to cancel a pesticide only if there exists reliable evidence regarding risk, exposure, and benefits. These issues may be addressed in 1991 when Congress considers the FIFRA reauthorization bill.

Reregistration has an unglamorous but vital role in the pesticide program. Its successes are little noted; its failures are heavily criticized. The desirable result of a reregistration review is knowing enough about a pesticide to conclude that its use is not harmful and that it is properly labeled. Such good news is, of course, no news. But if reregistration doesn't go as fast or as well as people think it should, as is almost inevitable, that's big news.

Over the past 20 years we have all learned a great deal about how difficult the job is. We now know much, much more about pesticide safety than we did previously, largely as a result of the steady progress of the reregistration program. We are reaching the point when the inevitably continuing debate about the wisdom of pesticide use will take place against a good background of modern information on toxicity, exposure, and benefits associated with individual pesticides. If policy debates about pesticides are based on knowledge rather than on fear of the unknown, the reregistration program will have served us well.

References

1. Federal Insecticide, Fungicide, and Rodenticide Act, *U.S. Code* Title 7, Sections 136–136y, 1990 ed.

2. Public Law 80–104, 61 Stat. 163 (June 25, 1947). That law required for the first time that products be registered by the U.S. Department of Agriculture (USDA) before they could be sold in interstate commerce. From then on, the law has required premarket *review* of each product by the USDA, although premarket *approval* was not required until the law was again amended in 1964. In addition, the 1947 FIFRA for the first time required that product labels must contain appropriate warning or caution statements and use directions. And the law said that a pesticide was misbranded if, when used in accordance with its labeling, it was "injurious to living man".

3. EPA was formed in December 1970; before then, FIFRA was administered by the USDA. The reorganization plan that formed EPA also transferred to EPA all the USDA personnel, funds, records, regulations, and physical assets associated with the administration of FIFRA.

4. For example, until FIFRA was amended in 1972, its registration requirement applied only to pesticides that were distributed or sold in interstate commerce; products that remained solely in intrastate commerce were regulated only by the individual states. Likewise, before the 1964 amendments a registration could be obtained "under protest" even if USDA did not think that the product met the statutory criteria; under the pre-1964 law, the ultimate function of a registration was to inform the government of what was being sold, not to preclude the sale of unsafe products. Finally, until 1972 the labeling of a pesticide was not binding on users; it was regarded as advice that should be followed.

5. One noteworthy development is the increased concern about the effects of pesticides on wildlife, in particular the passage of the Endangered Species Act and its application to pesticide regulation.

6. Old products otherwise would have unfair and harmful advantages over new competing products. Manufacturers of old products could avoid both the costs of new types of testing and the risk of unfavorable test results.

7. House Committee on Agriculture, Report to Accompany H.R. 1237, H.R. Rep. No. 313, 80th Congress, 1st Session, April 23, 1947. The Senate committee report incorporated the House report verbatim.

8. A quite informal procedure would have been appropriate. Neither the 1947 FIFRA itself nor the Administrative Procedure Act (APA), enacted in 1946, required a formal hearing or a decision based strictly on a record made at such a hearing. The APA merely required that before a registration may be canceled, the registrant

must be given written notice of the basis for the revocation and an opportunity to demonstrate or achieve compliance with the requirements.

9. See *Pax Company v. United States,* 454 F. 2d 93 (10th Cir. 1972). In this case the USDA had informed the registrant that an arsenic-containing product would not be reregistered because of the risks to human health.

10. Public Law 88–305, 78 Stat. 190 (May 12, 1964); *U.S. Code* Title 7, Sections 135–135k, 1970 ed.

11. *Stearns Electric Paste Co. v. EPA,* 461 F. 2d 293 (7th Cir. 1972).

12. Public Law 92–516, 86 Stat. 973 (Oct. 21, 1972); *U.S. Code* Title 7, Sections 136–136y, 1974 ed.

13. *Federal Register* 40 (July 3, 1975) p 242.

14. *Federal Register* 44 (December 26, 1975) p 76311.

15. *Federal Register* 53 (May 4, 1988) p 15952.

Chapter 6

Pesticide Regulation in Developing Countries of the Asia–Pacific Region

Edwin L. Johnson

The use of pesticides (1) in the developing world has been increasing over the past decade. The rate of growth in pesticide use has increased perhaps faster in the Asia–Pacific region than in any other region. Of course, even within the region, there are large differences in the rate of growth in pesticide use and in the numbers and kinds of pesticides in common use. Figure 1 presents rates of increase in the use of pesticides in several countries in this region. During the period 1980–1985, the average annual market growth was approximately 5–7%. However, during the same period, the market in Indonesia and Pakistan grew at the much faster rate of 20–30% each year.

These trends appear to have continued in terms of general magnitude since 1985. By way of contrast, the world market is estimated to grow at a rate of some 4.5% per year.

At the other end of the spectrum are some of the Pacific Island countries producing subsistence crops. The extreme opposite end of the spectrum is Wallis and Futuna Islands, which use only a few pesticides, all of which are purchased and applied by government agencies. However, as some of these island countries develop cash crops for export, the use of pesticides can be expected to increase. This trend has already been demonstrated by events in countries such as Western Samoa and Tonga.

The population of the Asia–Pacific region is expected to double between 1980 and 2000. If this population is to be fed,

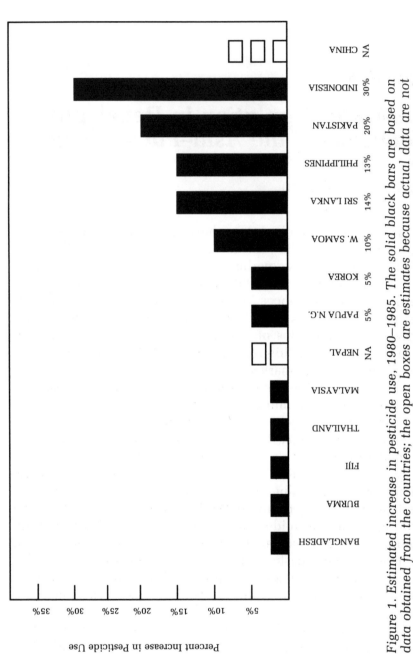

Figure 1. Estimated increase in pesticide use, 1980–1985. The solid black bars are based on data obtained from the countries; the open boxes are estimates because actual data are not available. (Reproduced with permission from ref. 2. Copyright 1987 Asian Development Bank.)

the production of food supplies can be expected to grow at a rate at least equal to the rate of population increase. Considering increased per capita income in the developing countries that are also industrializing countries, and the search for increased exports, food production rates will be even higher than population growth. An assessment of the available land resources suggests that 75% of the extra food should come from higher yields. In turn, these yields are likely to require major crop intensification programs together with major increases in inputs including pesticides (2).

The Asia–Pacific region is predominantly an insecticide market. Of the total estimated consumption of pesticides, about 75% are insecticides, 13% are herbicides, and 8% are fungicides. Insecticides are used mainly on rice, cotton, and vegetables; herbicides on rubber, palm oil, tea, coffee, and cacao plantations; and fungicides for control of diseases in tobacco, vegetables, and bananas.

This increase in pesticide use has been the source of considerable concern for a variety of reasons. First, some of the more toxic compounds have come into use with cessation of use of the older organochlorines by farmers who are untrained in application techniques and lack protective equipment. Second, absence of effective regulatory programs results in poor labeling and quality control of products, inadequate training of applicators and health practitioners, ineffective enforcement and lack of monitoring of exposure, environmental burdens and effects, and food residues.

U.N. Food and Agriculture Organization International Code of Conduct

Early in the 1980s, alarms raised by various groups about this variety of real and perceived pesticide problems led a number of governments, international organizations, the pesticide industry, and environmental groups to intensify efforts to find ways and means to reduce the problem. But each of these groups offered different solutions based on their different interests and responsibilities. Furthermore, a number of governments and organizations questioned the propriety of supplying pesticides to countries that did not have adequate institutional capacity to ensure the safe application and use of pesticides.

In this context, in autumn 1981, the United Nations Food and Agriculture Organization (U.N. FAO) launched the idea of an "International Code of Conduct on the Distribution and Use of Pesticides". In consultation with appropriate United Nations and other organizations, the code was developed and finally adopted by all FAO member nations in 1985 (3).

The code provides a uniformly approved statement of ethical behavior for all parties involved in pesticide distribution and use—importing and exporting country governments, international organizations, industry, public interest groups, and users. Although the code is voluntary, it provides a common standard of ethical behavior against which actions by these parties may be assessed. And, for purposes of this discussion, it provides a basis for evaluating the state of pesticide regulation and regulatory problems in the Asia–Pacific region.

The code consists of 12 articles. The title of the sections give an idea of the broad scope of coverage of this internationally approved code of behavior:

- objectives
- definitions
- pesticide management
- testing of pesticides
- reducing health hazards
- regulatory and technical requirements
- availability and use
- distribution and trade
- information exchange
- labeling, packaging, storage, and disposal
- advertising
- monitoring observance

Regulatory Status

The status of regulation in the Asia–Pacific region is reviewed here in terms of government responses to an FAO questionnaire distributed in December 1986 to all U.N.-member governments. This information represents a baseline against which future progress may be measured. Figures 2–9 (*4*) show the responses by governments in the Asia–Pacific region with respect to selected questions that relate to the governments of countries importing pesticides under the code (*5*). By and large the countries of the region may be considered primarily importers of pesticides, although there is some amount of manufacturing and formulation.

The FAO questionnaire consisted of 121 questions arrayed by article of the code. Trying to present all of them would be both cumbersome and confusing; therefore, I have followed a method developed by Hicks and Mobray (*6*) for questionnaire data from the Pacific and applied it to Asia as well. The approach makes a comparison on eight different aspects that cover the main themes of the code rather than on the specific articles. It groups questions that collectively give a picture of the status in the region with respect to government responsibilities. The themes are

- need for legislation
- legislation
- pest management
- technical advice–assistance
- health and safety
- packaging
- labeling
- advertising

For additional simplification, in cases where the FAO questionnaire allowed for answers between a simple yes or no, for example, "partially" or "sometimes", all responses other than yes are grouped together as no. This practice tends to present a slightly more negative picture with respect to the degree of compliance with the code than might otherwise be the case.

For purposes of comparison the percentage of countries in Africa and in the so-called developed countries is also presented.

Various measures of legislative and regulatory development are shown in Figure 2. About 65% of the countries in the region report that they observe the code, and only about 15% reported that use of pesticides is not legislated. However, substantial per-

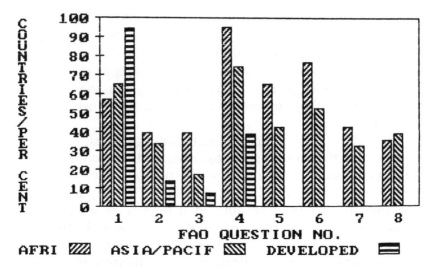

Figure 2. Responses to the December 1986 FAO questionnaire of governments in Africa, the Asia–Pacific region, and the developed countries. The question related to legislation. Key: 1, the country observes the FAO code; 2, distribution is not regulated; 3, pesticide use is not legislated; 4, government resources are not adequate; 5, there is no pesticide registration or control scheme; 6, there is no pesticide control legislation; 7, there are no limitations on availability; and 8, the government cannot enforce highly toxic pesticide restrictions.

centages of countries reported the absence of pesticide registration schemes and pesticide control legislation.

More than 30% reported that there are no limitations on availability or that they cannot enforce restrictions on highly toxic pesticides. No developed countries reported the absence of these key elements of pesticide regulation. About 75% reported inadequate resources, which can sometimes be helped by the presence of legislative requirements that focus funding decisions or provide for the charging of fees to support the program.

Despite the absence of legislation and regulations, one can still ask whether these are actually needed to address problems. For example, with low pesticide use, legislation may not be absolutely necessary. Figure 3 and most of the following figures shed some light on the need for legislation and regulatory programs. More than 40% of the countries reported unsatisfactory

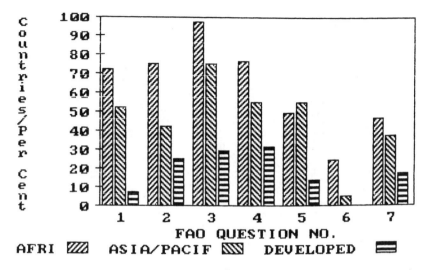

Figure 3. Responses to the December 1986 FAO questionnaire of governments in Africa, the Asia–Pacific region, and the developed countries. The question related to the need for legislation. Key: 1, industry does not act responsibly; 2, export trading is not satisfactory; 3, traders do not follow products to users; 4, there is no review of pesticides marketed; 5, the sale of unsafe products is not stopped by industry; 6, industry does not help correct problems; and 7, export quality is not adequate.

conditions in a variety of areas shown in Figure 3. Comparing this result with the situation in developed countries that generally have pesticide regulatory legislation and regulation clearly demonstrates a need for such authorities in the countries of the Asia–Pacific region. This conclusion is not meant to imply that they all need the same level of regulation as the developed countries or that every country in the region needs the same level of regulatory control. This level of regulation must be tailored to the situation in each country.

Most countries in the region have participated in several workshops given by the FAO, the Asian Development Bank, the World Bank, the U.S. Agency for International Development, the U.S. Environmental Protection Agency (EPA), and others, either jointly or separately, and have noted that the assistance has been useful in improving their regulatory capability. In addition, the FAO project to implement the FAO code in the region has provided significant input to the countries. The responses in Figure

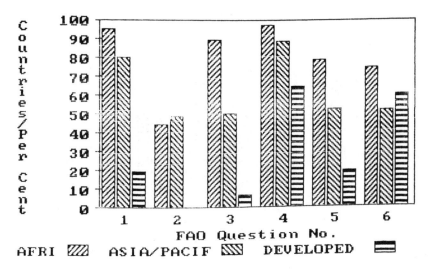

Figure 4. Responses to the December 1986 FAO questionnaire of governments in Africa, the Asia–Pacific region, and the developed countries. The question related to technical assistance or advice. Key: 1, there is insufficient educational material; 2, export notifications are not received; 3, analytical methods are not provided when requested; 4, no help is given to train analytical staff; 5, no local facility is available to verify quality; and 6, no assistance is available to train data evaluators.

4 indicate that this input is important. The activities cited as examples of such assistance have largely occurred since the date of the FAO questionnaire; therefore, it is hoped that these activities are helping to meet the needs of these countries.

Of particular interest is that some nations reported that they did not receive export notices for banned or severely restricted pesticides. Principles for such notification were adopted by the Organization for Economic Cooperation and Development (OECD) and the United Nations Environment Programme (UNEP) in the early 1980s and in the FAO code in 1985. A recent review by the OECD indicates that only a few OECD countries are actually implementing these procedures. This finding is reaffirmed more generally for member nations of the FAO in responses to the questionnaire. Apparently, exporting countries need to make significant improvements in their implementation of the notification system provisions of these internationally approved procedures.

Despite the emphasis placed on integrated pest management (IPM) by international technical organizations, lending institutions, and donor agencies, Figure 5 indicates very little promotion of these practices in the region, nor is there pursuit of other activities related to development of national pesticide strategies related to IPM. There are, of course, specific examples of introduction of major IPM efforts in many of the countries of the region, but the responses to the FAO questionnaire do not show this effort to be major. In the absence of significant control of pesticide use as noted earlier, it is not easy to promote these programs. An extensive extension activity would also be needed.

Figure 6 shows a large number of reported health and safety problems. Interestingly, more developed countries reported the availability of very toxic pesticides and pesticides stored in food outlets than did countries in the Asia–Pacific region. However, in response to other questions, these countries also reported the ability to control availability and use of such pesticides and that

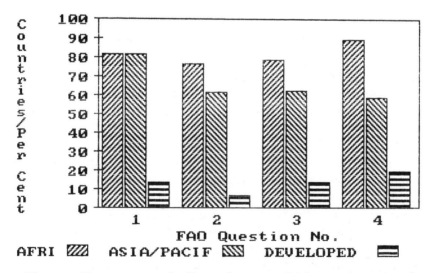

Figure 5. Responses to the December 1986 FAO questionnaire of governments in Africa, the Asia–Pacific region, and the developed countries. The question related to pest management. Key: 1, integrated pest management is not promoted; 2, resistance strategies are not developed; 3, environmental effects are not studied; and 4, use–import statistics are not collected.

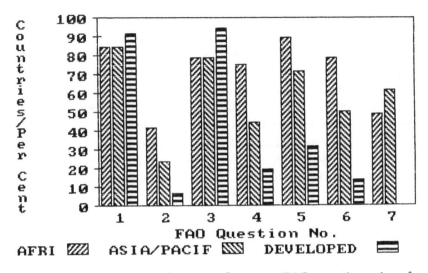

Figure 6. Responses to the December 1986 FAO questionnaire of governments in Africa, the Asia–Pacific region, and the developed countries. The question related to health and safety. Key: 1, very toxic pesticides are available; 2, no effort to provide less toxic pesticides; 3, pesticides are stored in food outlets; 4, pesticides are not adequately segregated; 5, the training of salesmen is inadequate; 6, users are not adequately informed; and 7, no guidance on poisoning treatments is given.

adequate segregation of pesticides from foods is achieved. Availability and lack of control of access and use, protective measures, and training in safe use and handling are important in combination.

Further indications of the need for increased regulation of pesticides in the region are the reported problems in labeling (Figure 7), packaging (Figure 8), and advertising (Figure 9). Responses shown in Figures 7 and 8 show that the presence of regulatory programs leads to more satisfaction with the labeling and packaging of pesticides. Even countries with developed regulatory programs do not appear to be more satisfied with advertising than do those with less sophisticated or even no pesticide legislation or regulation.

The FAO plans to repeat the questionnaire survey periodically to provide a picture of changes in the regulatory programs of countries and the implementation of the code. It will be en-

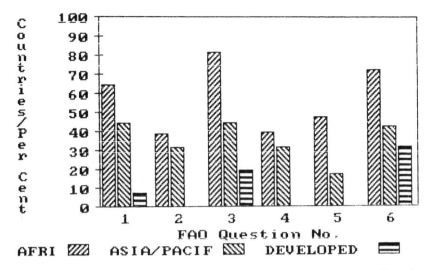

Figure 7. Responses to the December 1986 FAO questionnaire of governments in Africa, the Asia–Pacific region, and the developed countries. The question related to labeling. Key: 1, labeling is neither adequate nor effective; 2, careful and clear labeling is not required; 3, labels do not conform with official recommendations; 4, hazard classification is not shown on labels; 5, container reuse warnings are not given; and 6, no batch numbering is used.

lightening to compare the results of the next survey with this one to see the degree of progress being achieved.

Trends in the Asia–Pacific Region

Having reviewed the baseline situation in Asia and the Pacific, it is useful to look at the trends in pesticide regulation. This information is subjective, deriving from my participation in three workshops on strengthening pesticide regulation that were held in Bangkok in November 1988 (7), Noumea in March 1989 (6), and Manila in July 1989 (8, 9). Also considered are the activities of various institutions active in the region to implement the FAO Code of Conduct and their expected impact on the regulatory picture.

Most of the countries in the region are at some stage of establishing, changing, or implementing legislation and registra-

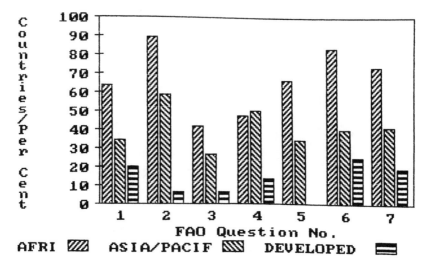

Figure 8. Responses to the December 1986 FAO questionnaire of governments in Africa, the Asia–Pacific region, and the developed countries. The question related to packaging. Key: 1, packages are not suitable; 2, good packaging practices are not used; 3, ready-to-use packages are not available; 4, childproof containers are not used; 5, there is no provision to stop repackaging; 6, repackaging provisions cannot be enforced; and 7, appropriate size packs are not available.

tion programs. Countries lacking basic laws and regulations are developing them. Countries more advanced in pesticide regulation, such as the Philippines and Malaysia, are expanding programs into areas of worker protection and monitoring. Tonga is implementing a new law. Nepal is developing basic legislation. Most likely, all countries in the region will have basic regulatory structures in place in the next few years.

The countries in the region can be expected to increase communication and networking among themselves and with developed countries and international organizations as a result of various efforts in the region (to be described later). This increased communication should increase their capability to deal with health, safety, and environmental issues in the absence of adequate staff and to support each other in reaching regulatory decisions. The Pacific Islands may develop an approach to pesticide approval that makes use of actions in other countries in

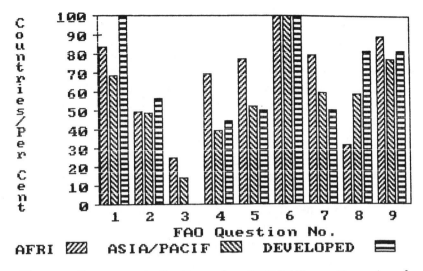

Figure 9. Responses to the December 1986 FAO questionnaire of governments in Africa, the Asia–Pacific region, and the developed countries. The question related to advertising. Key: 1, unsubstantiated claims are a problem; 2, misleading advertising is encountered; 3, confusing distinguishing names are used; 4, nonrecommended uses are promoted; 5, product safety claims are made; 6, misleading statements are made about profits; 7, advertisements do not encourage label reading; 8, complaints are received about advertising; and 9, safe and effective use practices are not advertised.

the region that have more intensive regulatory programs, such as Australia and New Zealand. Therefore, registrations in those two countries may become more important to manufacturers wishing to market in the Pacific.

One of the principal problems identified is the lack of analytical capability to ensure the quality of products marketed in the region. These countries for the most part probably will not be able to establish and operate such laboratories. This problem becomes more serious as the number of "me-too" products* increases and some governments press the idea of generic registration to decrease the cost of pesticides to farmers. (The concept of generic registration was raised at the Manila workshop as a serious proposal in at least three countries and is of concern to

---

*A me-too product is one that is represented as equivalent or substantially similar to a previously registered product.

those countries' regulators. The December 1989 meeting of the FAO Expert Group on Pesticide Registration addressed this issue and recommended that governments not adopt generic registration except as it conforms to "me-too" registration in the FAO guidelines. This recommendation means that detailed information on specifications and quality control are necessary requisites.)

Apparently, countries in the region will be trying to reduce the use of World Health Organization (WHO) Hazard Category 1A and 1B pesticides. Training large numbers of small farmers requires significant resources. Thus, some countries are banning very hazardous pesticides when less hazardous pesticides, formulations, or packaging become available.

Many countries in the region have expressed the desire to know more about exposure of workers to pesticides and will, through exchange of information and exposure monitoring assistance from organizations such as WHO and FAO, develop a better understanding of this issue. The absence of protective clothing that is both affordable and practical in hot climates may lead to the stronger regulation of some pesticides.

Harmonization of registration data requirements and use of external data will likely increase. The implementation of the Prior Informed Consent (PIC) procedures in the FAO code, formally approved by member governments in November 1989 at the FAO Conference, and also approved by the UNEP Governing Council, should result in fuller information exchange. Absence of such information is a subject of complaint by these countries, as already noted. Information exchange, coupled with the requirement of PIC for exporting countries to ensure compliance with decisions of importing countries, can be expected to tighten regulatory controls. This expectation is particularly true in light of the support indicated for these procedures by the international pesticide industry represented by GIFAP (International Group of National Associations of Manufacturers of Agrochemical Products). Certain pesticides in the WHO Hazard Classification 1A will be included in the PIC procedure as well as those that have been banned or severely restricted for health or environmental reasons.

Important sources of data review and assessment that could be of value to developing countries include the Health and Criteria Documents of the WHO International Programme on Chemical Safety and the data available on toxicology and residues from the FAO–WHO Joint Meeting on Pesticide Residues. Such data and information from reports of these expert groups could improve harmonization of regulatory approaches and reduce the resource needs of individual countries.

Driving Forces in Asia and the Pacific

The most significant stimulus is the realization by the countries themselves that more regulation of pesticides is needed with increasing use, with use of highly toxic products, and with increasing imports of commodity pesticides as patents expire.

In addition, other stimuli in the region are the several international organizations including FAO, the Asian Development Bank, the World Bank, and the Pesticide Action Network/International Organization of Consumer Unions (PAN/IOCU).

The FAO, operating from its Bangkok-based project office, is conducting a multiyear program to implement the Code of Conduct in Asia and the Pacific. This project is funded by a grant from the Government of Japan. Its purposes are to ensure that pesticides are used only under good agricultural practices, in a safe, efficient, and environmentally sound manner. This project provides technical assistance and training to individual countries. It has provided consultants to assist in drafting laws and regulations and upgrading registration and post-registration activities.

The Asian Development Bank and the World Bank are paying greater attention to pesticides procured under agricultural input loans. The types of pesticides to be used are scrutinized in the light of the country's capability to reduce the risk of using inappropriate materials. More important, technical assistance components are often added to such loan agreements to require the recipient country to improve its regulatory capability. The banks have also funded multicountry training.

Also, bilateral agencies such as the German Agency for Technical Cooperation (GTZ) and the U.S. Agency for International Development provide assistance to countries on activities that strengthen regulatory capability. An example is the Malaysian–German Pesticide Project, which has assisted in development of regulatory procedures, engaged consultants, and provided training both in-country and abroad for Malaysian regulators.

The U.S. EPA has provided, and will continue to provide, technical expertise and information and worked in collaboration with FAO and the other governmental and international bodies in strengthening pesticide regulation.

The international pesticide industry (GIFAP) is requiring member associations to adopt the code responsibilities with respect to their activities and have been active in such areas as providing technical expertise, developing training and instructional materials, and researching protective equipment suitable to tropical agriculture.

Finally, the public interest nongovernmental organizations, principally PAN/IOCU, are active in the region, bringing infractions of the code and pesticide problems to the attention of regulators in individual countries. This informing is often done through the media and the political process. This public involvement will continue to bring pressure on the regulatory authorities and processes and is growing in intensity.

Summary

Although there are great differences among the countries in the region in terms of specific pesticide issues and stage of regulatory development, the Asia–Pacific region as a whole has achieved greater progress in strengthening pesticide regulation through implementation of the code than any other region of the world. This progress has been accomplished in part through the active support of a number of international, multilateral, and bilateral organizations operating in the region, but mainly because of the realization of the countries that action is necessary. The concern and initiative and receptivity of the countries of the region has been an essential component.

Despite this progress, the countries in the region generally agree that much more needs to be done in the regulation of pesticides. Through sustained national efforts and regional cooperation, a much more effective implementation of the code will be attained in the next few years. This hope is supported by the rapid progress being made in the past year by countries in the development and improvement of laws, regulations, and operation of programs and networking among countries.

References

1. The term "pesticides" is defined by the FAO International Code of Conduct: "Pesticide means any substance or mixture of substances intended for preventing, destroying, or controlling any pest, including vectors of human or animal disease, unwanted species of plants or animals causing harm or otherwise interfering with the production, processing, storage, transport, or marketing of food, agricultural commodities, wood and wood products, or animal feedstuffs, or which may be administered to animals for the control of insects, arachnids, or other pests in or on their bodies. The term includes substances intended for use as a plant-growth regulator, or agent for thinning fruit or preventing the premature fall of fruit, and substances applied to crops either before or after harvest to protect the commodity from deterioration during storage and transport." (2) Individual governments may define the term differently; however, almost all include the chemicals and biological agents of major use in agriculture, forestry, and vector control, including insecticides, herbicides, fungicides, and rodenticides. Many governments regulate all but vector control pesticides in the Ministry of Agriculture and vector control products in the Ministry of Health.
2. *Handbook on the Use of Pesticides in the Asia–Pacific Region*; Asian Development Bank: Manila, Philippines, November 1987.
3. *International Code of Conduct on the Distribution and Use of Pesticides*; FAO: Rome, 1985.
4. In Figures 2–9, the percentages reported are the percent of governments answering the question as stated divided by the total number of governments responding to the question. Because all governments did not respond to all questions, the total may vary, although most responded to the questions noted in these figures. Some governments did not respond to the questionnaire at all, and they are unaccounted for in this analysis. Details on numbers of respondents for all questions can be found in reference 5.
5. *International Code of Conduct on the Distribution and Use of Pesticides: Analysis of Responses to the Questionnaire by Governments*; FAO: Rome, January 1989.
6. *Report of the Regional Workshop on the Implementation of the FAO Code of Conduct on the Distribution and Use of Pesticides for the*

Pacific Islands; FAO/SPA: Noumea, New Caledonia, March 6–10, 1989.

7. *Report of the Workshop on Pesticide Regulatory Principles and Procedures for the Asia and Pacific Region*; FAO: Bangkok, November 14–25, 1988.

8. Draft Report of the Regional Workshop and Symposium on Strengthening Pesticide Regulations, Asian Development Bank: Manila, Philippines, June 26–July 7, 1989.

9. Gaston, C. G., "Progress in the Implementation of the International Code of Conduct on the Distribution and Use of Pesticides in the Asia and Pacific Region", unpublished, presented at the Regional Workshop and Symposium on Strengthening Pesticide Regulations sponsored by Asian Development Bank, FAO et al., Manila, Philippines, June 26–July 7, 1989.

Chapter 7

Pesticide Registration in Europe

Barry Thomas

In Europe over the past 20–30 years, sophisticated and extensive governmental systems have been developed to assess the safety and efficacy of pesticides before they are approved for use. During this period data requirements to support registration have become more demanding, not only to reflect parallel scientific developments in risk assessment techniques, but also in response to increasing public and political pressure for greater reassurances regarding the safety of pesticides, pharmaceuticals, and chemicals in general. More recently a "greening" of the political climate has led to the realization by parties of differing political persuasion that votes are to be won by being seen to be "environmentally friendly". This objective in turn has manifested itself in an overall "hardening" of official attitudes towards pesticide usage, resulting in increasing data requirements and making it more difficult to achieve registration of new pesticides or the reregistration of existing products.

Although a certain degree of uniformity exists between the data requirements of the European regulatory authorities, particularly in the area of toxicology, major differences, especially in environmental data requirements, are also encountered. This chapter will concentrate on these differences and also on the attempts by certain European regulatory authorities to introduce policies aimed at the overall reduction of pesticide usage. Additionally, and in a sense in contrast, the proposals from the European Commission to harmonize pesticide registration by introducing common regulatory procedures and common data requirements within the European Community will be discussed.

European Data Requirements

The basic data requirements (1) for a new pesticide are both comprehensive and extensive and cover the areas of chemistry, toxicology, plant and animal metabolism, residues, environmental fate, and environmental impact. The details are well known and will not be repeated here, but additional studies will very often be required to reflect the specific properties or proposed use of the pesticide in question. Against the background of these basic data requirements, the major European countries have additional and individual requirements, the key elements of which can be summarized as follows.

As has already been mentioned, the toxicological and residue data requirements needed to support registration have, to a large extent, remained fairly constant over the past few years, a fact that reflects the "state of the art" and the degree of harmonization achieved in these two areas of data requirements. In contrast, environmental data requirements have increased significantly, and there is every indication that this increase will continue into the foreseeable future. These increased requirements are typically characterized by the need for additional data on the behavior of pesticides in soil and water, their toxicity to a wider range of aquatic animals, both on an acute and longer-term basis, effects on soil organisms and soil processes, etc. Additionally the need for a more comprehensive understanding of the metabolism in plants and animals, and particularly domestic animals, has become more of a "standard" requirement than was previously the case.

These additional requirements have most clearly been demonstrated in such territories as Germany, Denmark, the Netherlands, the United Kingdom, and Ireland, although, as will be discussed, this list of territories is unlikely to remain exclusive. It is interesting to compare the costs of the additional data requirements of individual European countries with the value of these markets (2) (Table I). These differences in market value per pound of additional cost reflect differences in two major factors, namely, the market size of the individual territory and the cost of the additional data required to obtain registration in that territory. Thus, the "best value" is obtained in the largest markets such as France, Italy, and the United Kingdom. Con-

**Table I. Costs of Additional Data Needed for Registration and Value
of Pesticide Markets in European Countries**

Country	Cost (thousand)	Market Value (million)	Market Value per Pound of Cost
Belgium	10	80	8000
Denmark	105	100	952
Eire	10	30	3000
France	15	1000	66,600
Germany	140	400	2850
Greece	10	80	8000
Netherlands	40	120	3000
Italy	21	400	19,000
Portugal	10	40	4000
Spain	10	210	21,000
United Kingdom	15	380	25,300

All values shown are in pounds sterling and are not absolute.

versely, the "worst value" is encountered in those territories where the market is relatively small, but where the cost of additional data is high, for example, Denmark and the Netherlands. Where the intended use of a product is restricted to a small individual segment of the market, the commercial justification for attempting to register such a product may thus be borderline. The comparison between Germany, Italy, and the United Kingdom, which have similar market values, serves as a graphic illustration of the additional data necessary to obtain registration in Germany.

Current and Future Regulatory Trends

In Europe a number of major factors are either operative or are likely to gain momentum and thereby influence the regulatory control of pesticides.

1. Fixed Criteria The "registrability" of a pesticide judged against the yardstick of fixed criteria is most advanced in Denmark where such criteria are being used not only as a basis for the registration of new pesticides but also for the reregistration of existing products. Products that are judged to be "dangerous"

to health or to the environment cannot be registered. Criteria
have been established for

- acute toxicity
- subacute toxicity
- chronic toxicity
- carcinogenicity
- mutagenicity
- reproductive toxicity
- neurotoxicity
- soil persistence
- soil mobility
- bioaccumulation

More recently in the Netherlands, if the Environmental Criteria
that have been proposed (3) had been implemented per se, the
result would have been an estimated withdrawal of between
75% and 90% of products currently marketed.

2. Safer Alternatives The basic philosophy of this approach
is that a pesticide will not be registered if products that are currently acceptable for the same purpose are presumed to be less
dangerous to health or less harmful to the environment. This
policy is most well developed in Denmark and Sweden, and was
also included in the overall Environmental Criteria proposed in
the Netherlands. The implications of this policy on attempting
to introduce new active ingredients are self-evident.

3. Reduction in Pesticide Use A number of countries have
declared their intention to reduce the overall use of pesticides
(4). At present these are Sweden, Denmark, and the Netherlands,
and various targets have been set. For example, in both Sweden
and Denmark the objective is a 50% reduction to be achieved
by a number of measures, including

- deregistration of "dangerous" pesticides
- use of more desirable alternatives
- more efficient application
- lower dose rates
- fewer applications
- integrated pest control

Similarly, the European Commission, in its communication "Environment and Agriculture" (5), has declared as a policy objective that throughout the European Community the aim should be to "reduce to a strict minimum the use of chemicals for agriculture".

4. Reregistration of Existing Products Reregistration can, to a large extent, be seen as a means by which the preceding three factors can be implemented. Invariably such reregistration exercises are accompanied by requests for additional data that may often be satisfied only by a major research program. Such reregistration exercises are currently being undertaken in Denmark, Germany, and the United Kingdom.

The European Commission Registration Directive

In contrast to the diversity of requirements and policies being implemented in some European countries, the European Commission published its proposals to harmonize pesticide registration throughout the European Community (6). The major elements of these proposals can be summarized as follows:

1. A Community procedure will be used for evaluating active ingredients in accordance with defined data requirements.

2. Active ingredients found to be acceptable will be included on a Community "positive" list.

3. Registration of formulations will be the responsibility of individual member states, but only those formulations containing active ingredients on the positive list may be registered.

4. Existing active ingredients (an estimated 430 are currently marketed in at least one member state) will be reviewed by the same Community procedure within a 10-year period following implementation of the Directive.

Careful consideration of these proposals clearly indicates that, to a large extent and in any practical sense, the degree of harmonization achieved is minimal. Member states will be allowed to control pesticide registration much as they do now; existing national differences in data requirements will be left more or less intact. The Draft Directive, as published, concentrates on harmonization of data to support the active ingredient, but during discussions at the Council, Working Party proposals have also emerged for extensive data requirements for the formulated products. These data requirements will be supported by Uniform Principles that will give guidance on when data are required, which protocols are to be used, the use of models, and data interpretation.

The final format of the Directive remains somewhat uncertain until its publication, which is anticipated during the second quarter of 1991. Work has also started on drafting the Uniform Principles that will be introduced into the Community as a second Directive, the implementation of which will coincide with that of the main Registration Directive in mid-1993. Additionally, proposals are being discussed regarding the procedures for reviewing existing active ingredients. These proposals are likely to involve the identification of five priority groups comprising approximately 90 active ingredients per group followed by reviews undertaken by individual member states.

Detailed operation of this new European registration scheme remains to be defined, but the impetus of the Single European Act and the establishment of the open market by 1992 cannot be overlooked in estimating the political will to achieve the overall objective of a Community system for the control of pesticides.

Conclusions

Many European countries are demanding substantial data, particularly environmental data, in support of current and new registrations in an unprecedented way. Despite the likelihood of a European Registration Directive within the next 2 years or so, these demands may well continue. The consequence to industry is that new product registration will undoubtedly be more difficult to attain, not least in terms of the time taken to achieve such registration. Additionally, many, if not all, of a company's strategic products will require support to meet these demands. Without such support some products will not be able to remain on the market, and inevitably many minor products will have to be withdrawn because the resources needed to sustain them cannot be commercially justified.

References

1. *Pesticides;* 6th ed.; Council of Europe: Strasbourg, France, 1984.
2. Foulkes, D. M. "Pesticide Regulation: Perception and Reality", In *Pesticide Outlook*; Stell, G., Ed.; The Royal Society of Chemistry: Cambridge, England, 1989; Vol. 1, pp 19–22.
3. "Long-Range Crop Protection Plan", Ministry of Agriculture, Nature Management and Fisheries, The Hague, Netherlands, August 1990.
4. Bernson, V. "Regulation of Pesticides in Sweden", *Proceedings: 1988 British Crop Protection Conference; Pests and Diseases*; BCPC Publications, Thornton Health: Surrey, England, 1988; pp 1059–1064.
5. "Environment and Agriculture", European Commission Communication, Commission Document COM(88)338 Final, June 8, 1988.
6. "Amended Proposal for a Council Directive Concerning the Placing of EEC-Accepted Plant Protection Products on the Market", EC Directive COM(89)34 Final, *Official Journal* No. C89, Luxembourg, Belgium, April 10, 1989; pp 22–40.

Chapter 8

Academic and Government Research Input to the Registration Process

G. Wayne Ivie

Essentially all scientific research on agrochemicals is done by individuals employed by either the agrochemicals industry, by various state entities—mainly the universities—or by the Federal government. And, although industry continues to take the lead role in agrochemical development as it has in the past, academic and government workers have historically had a significant research role in the process. My comments are centered on the historical and evolving relationships between the agrochemicals industry and the public sector over the past 20 years, specifically as regards the development and registration of pest management chemicals.

First a disclaimer: I make no pretenses of having a sufficient long-term perspective on the agrochemicals industry or on the academic and government liaisons with that industry to give a complete or fully accurate analysis of historical relationships and evolutionary trends. However, I did spend 6 years in academia pursuing two graduate degrees and doing postdoctoral work—all in pesticide research, and I'm now finishing my 18th year with the Federal government in a laboratory surrounded by the campus of a major land grant university, again doing primarily agrochemicals work. All these years, I've maintained a reasonably close and productive liaison with my government, academic, and industry colleagues, and I've noticed over time a distinct evolution in the interrelationships between these three components of the agrochemicals research community. I don't

claim that my perspectives on this subject are totally right, but
I do believe they are mostly right, and I offer them for what
they're worth, perhaps out of nostalgia, perhaps to satisfy some
need I see for a historical recording, but mostly I hope to stim-
ulate thought, analysis, and introspection.

The Cooperative Spirit

During 1966–1972, the time of my graduate and postgraduate
work, the agrochemicals industry actually underwrote my scientific training. The radioisotopes I used in my metabolism
work with two new carbamate insecticides were simply not
available from any other source; they were provided to me by
industry at no cost, with good feelings, and with few strings
attached. Metabolite standards and other needed chemicals
were freely given, as was advice and counsel as needed. Limited
but sufficient financial support was provided, and I think it significant that my first trip to a national scientific meeting was
directly and totally funded by the supporting agrochemicals
firm. In my later graduate work on photosensitized decomposi-
tion of pesticides, at least half a dozen companies freely gave of
their sole-source chemicals (radiolabeled and otherwise), even
though it was clear that my work likely would have no direct
commercial impact. Representatives of one company I visited as
a Ph.D. student showed me their stock of pesticides and other
chemicals, told me to take what amounts I wanted of anything
they had, and then left the room!

Times were indeed simpler then, and my experiences during
that era were not unique. Industry, academia, and government
were in a very real sense all "in it" together, and the times were
exciting. This sort of freewheeling interaction continued for several years after I began my work as a Federal researcher—persisting at least through the late 1970s.

Why was this so? Almost simultaneous with the Division of
Pesticide Chemistry's formal birth 20 years ago, the U.S. Environmental Protection Agency (EPA) was created and assumed
pesticide registration responsibilities formerly held by the U.S.
Department of Agriculture (USDA). New and progressively more
detailed data requirements for registration quickly followed, and

demanded scientific expertise that was often not available to industry in-house. Industry sought and found highly qualified and willing cooperators in the public sector, and these scientists provided much-needed research data in a highly cost-effective manner. The perception in those early years—and it was probably valid—was that pesticide registration packages were strengthened by the contributions of academic and government researchers. The benefits went both ways. Public sector workers profited from these cooperative ventures through industry-provided financial or product support, from the opportunities associated with research on new compounds, and from the satisfaction of making very positive contributions to the pesticide registration process specifically, and to agricultural productivity generally. Equally significant (if not more so), through these relationships industry supported the training and development of highly competent agrochemical scientists. No doubt a great many Masters and Ph.D. degrees were earned during the 1960s and 1970s as a direct consequence of this cooperative spirit that existed toward agrochemicals development.

Pressures on Industry

This era began to draw to a close perhaps as early as the mid-1970s, but certainly no later than the beginning of the 1980s. At present, the level of such cooperative research interactions is probably not more than 10% of what it was 20 years ago. Why? In the first place, the regulatory forces that acted early in the era to pull the three groups together evolved in a way that ultimately began to push them apart. Concerns over the environmental and public health consequences of pesticides continued to mount during the 1970s, and the EPA quite rightly responded to the public pressures and congressional mandates by requiring industry to meet progressively more rigid and detailed criteria for successful registration. The cost to industry in putting together an acceptable registration package soared, and time frames for registration quickly lengthened to half a decade and more. Individual firms that were involved in pesticide development got committed for the long haul, or they got out! And to be successful and competitive, industry had to carefully plan and orchestrate the process. Studies had to be designed, scheduled, and conducted carefully, and meeting deadlines became a necessity rather than a fuzzy goal.

These evolving pressures and constraints faced by industry became less and less compatible with the relatively independent character of many academic and government researchers. Frankly put, we in the public research community were neither fully informed nor acceptably appreciative of the changing rules of the game. We continued to value the interactions and support, but we resisted the closer oversight, greater emphasis on staying within protocols, and increased criticality of deadlines. Many of us were, in a sense, "loose cannons", and industry learned quickly that to control the process, it had to control the appropriate expertise. And industry began to do just that; hiring scientists in critical areas rather than relying on those outside the company.

Actions of the Public Sector

The most significant failing of many of us in the public sector during this time was that we did not appreciate the extreme sensitivities that quickly evolved as part of the registration process. We continued to believe that once an agrochemicals firm had provided us with a couple of millicuries of radiolabeled compound, or with 25 grams of the technical product, we had the right to do anything with it that we wanted to. We didn't understand that surprises tended to stop the registration process in its tracks, often for months or even years, and often with absolutely staggering consequences to the affected compound and affected firm. Something so simple as showing a new pesticide to be toxic to an obscure nontarget species could freeze the process. It mattered not that the subject pesticide might have been the only one ever tested against that species, nor did it matter that the subject organism might ultimately be shown sensitive to a wide variety of chemicals, or that its populations could recover quickly and fully. The "bottom line" was that EPA had no choice but to view such data as indicating a potential negative impact, and red flags went up that were very difficult to lower. Many people can surely recall instances of such disasters during the 1970s. To this day, many in the public sector are not fully aware of the fact that an agrochemical in the regulatory spotlight is in an extremely sensitive and vulnerable position, and that any research on it that has potential environmental safety or public health ramifications must be done with a very careful eye toward fair play and proper perspective.

These, then, are the evolving forces that have led industry to an inevitable backing away from the public sector liaison in agrochemicals development. Other factors from the public sector side have reinforced that trend, particularly over the past 5 years or so. Other things being equal, academic researchers will by and large maintain an interest in industry collaboration so long as they receive appropriate financial support. Research in academia is "soft-money" driven to a very significant extent. The Federal agricultural research sector, on the other hand, is largely "hard-money" driven, and its scientists and managers generally do not view soft-money opportunities as viable determinants of program emphasis. In other words, dollars tend to drive programs in academia, whereas programs tend to drive dollars in the Federal sector. And in Federal laboratories such as my own, a gradual evolution has occurred over the past 20 years to the current philosophy that it is largely inappropriate to target Federal research dollars toward direct support of agrochemicals registration. One significant exception to this philosophy is with respect to research in support of registration of agrochemicals for minor uses.

The single overriding occurrence that led public sector scientists to lose enthusiasm for liaison with the agrochemicals industry was, in my opinion, the implementation of Good Laboratory Practices (GLP) regulations. Once GLP regulations were in place, it was to industry's clear advantage to comply both quickly and fully, and they did just that. The dollar commitments needed were soon made and, just as important, industry scientists rather quickly accepted that they had to fully endorse the concept. Public sector scientists, on the other hand, had a choice, and for many that choice was easy to make. A number of scientists in the public sector viewed GLP with skepticism, some fear, and as a threat to their intellectual and scientific freedoms. These concerns may not be valid, but they existed and still do. Public sector scientists began to make mental cost–benefit analyses as they contemplated research in areas that might be covered by GLP regulations, and that might therefore subject them to increased formalization of their work and to GLP data audits. Some public sector scientists are consciously avoiding certain research areas, and I see nothing on the horizon that will reverse this trend.

The Situation Today

The agrochemicals registration world of today is far different than it was two decades ago. In those earlier times, the public and private research sectors very much worked as a team, they were honest and open with each other, they drew on and complemented each other's scientific and intellectual strengths, and all saw and appreciated the ultimate goals. Today's situation is not a result of some combination of random events, but rather is a predictable consequence of the forces that have acted on the agrochemicals industry and the public sector research communities over the years. These forces include

- increased concern for the environment and public health
- the natural tendency of developed societies to require greater detail, increased accountability, and higher assurances
- the progression of the agrochemicals industry into one of higher stakes, higher risks, and higher rewards
- the inevitable philosophical conflicts between the more loosely structured public sector research community and that of an industry that must survive and prosper in a highly regulated and highly competitive world

In today's environment, many public sector scientists probably think it wise to stay out of the process as it applies to environmental safety or health-related research on agrochemicals that are in the spotlight of registration or reregistration. The stakes are high; the sensitivities are great; the spotlight is bright! At a minimum, those in academia or government who do choose to become thus involved must willingly accept that they are entering a different world in which they are open to a level of oversight, scrutiny, and even criticism that extends far beyond the normal peer review process.

What should the public sector role be as we enter the 1990s? Given where we are today and making reasonable extrapolations toward the future, most would probably agree that public sector scientists should not do for the agrochemicals industry what that industry can and by all rights should do for itself. Our role should be to complement but not substitute. In usual cases, the great amount of research and data generation required for registration or reregistration of a specific agrochemical should be conceived, orchestrated, and funded by the benefiting firm. Such is, of course, largely the case today and has been for several years.

Two circumstances exist in which it is to society's benefit that public sector researchers maintain an active and rather direct role in the agrochemical development and registration process. The first is in the generation of efficacy data. No single agrochemicals firm can possibly duplicate the wealth of scientific expertise and experimental capabilities that exist in the public sector and can directly be applied to biological activity evaluations. The second circumstance is in the area of what is commonly referred to as minor use. Industry simply cannot economically support in full the development of agrochemicals for use on commercially minor crop or animal species. The public sector has a valid role and responsibility in such cases. The industry–public sector interface that should identify minor use needs and follow through with ultimate submittals of registration packages should be working better than it is. Industry, of course, ultimately controls the process through its cooperativeness and even initiative, and should best view its role here as one of public service, even to the point of limited sacrifice. Many of the risks to industry that are associated with public sector interface, as discussed earlier, certainly apply in the minor use area. However, such risks can be managed successfully by effective communication that involves frank discussion of expectations, limitations, and sensitivities.

In addition, many research areas are usually not directly contributory to registrations of specific agrochemicals, but they are of major relevance to agrochemical development and regulation. Examples include basic research to define biological processes and chemical phenomena, the elucidation of mechanisms of biological action, and methodology development to enhance monitoring and regulatory activities. Quite obviously, public sector scientists should maintain an active and leading role in such areas.

Concerns for the Future

I close with two real concerns. The first relates to the dissemination of knowledge. Many people believe that agrochemicals research data are not reaching the literature as they once did. The reason is quite obvious: As a greater percentage of agrochemicals research is conducted by industry, a lesser percentage will ultimately be published. It is hoped that industry managers will be aware of the need for careful balance of legitimate corporate concerns with the broader need for information exchange and scientific advancement.

My second concern is that fewer and fewer young research professionals are choosing the agrochemicals sciences. I can accept that we are where we are out of a certain inevitability, but it is saddening to witness a progressive weakening of an interface that has been exceedingly effective in creating highly trained, highly motivated, and highly productive young professionals: scientists who could walk out of their academic laboratory on Friday, step into an industry laboratory on Monday, and "hit the ground running". I wonder about the future—will the agrochemicals industry be able to find the talent it needs to remain competitive as we move more and more into the global economy? I wonder if industry might already be seeing problems here—I can hardly believe otherwise. Solutions are not apparent to me, but the industrial community surely recognizes that a significant problem exists, that it will likely get worse, and that it will greatly affect their future. I hope that the agrochemicals industry will seek out and support creative approaches to ensure the availability of scientific talent that will so desperately be needed in the future.

Chapter 9

Trends in Agrochemical Formulations

Barrington Cross

The 1960s

In the 1960s, pesticidal formulations were relatively simple one-component active ingredient products. Both the physical properties, especially melting point and solubility characteristics, and chemical stability of the active ingredient played a key role in the choice of formulation type. The end-use of the product also strongly influenced the formulation decision. In general, products used for soil application were formulated as solids, whereas products used on leaves were liquids. The wettable powder formulation type was an exception to this generalization, because this solid-based-product is applied to the foliage by diluting and dispersing in water.

The two predominant liquid formulation classes were aqueous concentrates (AC) and emulsifiable concentrates (EC). Of these, the simplest is the aqueous concentrate, which is an aqueous solution of an active ingredient that may contain an adjuvant to assist leaf penetration and uptake. These formulations are environmentally sound and are nonflammable; therefore they pose no warehousing concerns. Many products were formulated as an aqueous concentrate by preparing water-soluble salts of the acid active ingredient. Examples include 2,4-D (2,4-dichlorophenoxyacetic acid salts) and dicamba (2-methoxy-3,6-dichlorobenzoic acid salts). To the present day, aqueous concentrates remain a first-choice formulation whenever the solubility characteristics in water and the resulting chemical stability profile will allow. Products introduced as aqueous concentrates in the 1970s include Avenge herbicide (difenzoquat) and in the 1980s (1) Scepter herbicide (imazaquin).

Emulsifiable concentrates (EC) contain an active ingredient with a surfactant dissolved in a water-immiscible solvent, which upon dilution with water gives an emulsion for end-use application. The EC was the most widespread type of formulation used in the 1960s and remains so. Organic pesticides readily soluble in lipophilic solvents such as xylenes or combinations of chlorinated benzenes, toluenes, and xylenes are ideal choices for EC products. The dinitroaniline herbicides are invariably formulated in this way. Trifluralin is a product that can be formulated in monochlorotoluenes as a 4- or 5-lb/gallon EC product.

Solid-based granular formulations (C) can be formulated with sorptive or nonsorptive carriers. The sorptive granular products contain up to 25% of active ingredients, whereas the nonsorptive sand or limestone products are more limited in loading capacity, 5% active ingredient being most practical. The discovery of a wide range of organophosphate insecticides (2) in the late 1960s led to a series of products formulated on clays such as montmorillonite or attapulgite. The insecticide phorate, formulated as Thimet 20G, is such a product.

Wettable powder (WP) formulations represent the third type of solid product. The active ingredient, as a solid, is either absorbed onto or into a water-dispersible powder. Upon water dilution, the active ingredient is dispersed for application. Fungicides such as benomyl were most frequently used as WP products. The WP is in declining use in the United States because of the potential for undesirable dusting during dilution prior to application. However, water-soluble bags containing WP products are gaining in importance and overcome the dusting issue.

The active ingredient's properties (physical and chemical), the state of technology (AC, EC, C, or WP), the competitors, and the marketplace (the farmer) formed the microenvironment that influenced pesticidal formulations. The macroenvironment included the public and the government regulatory agencies (Figure 1).

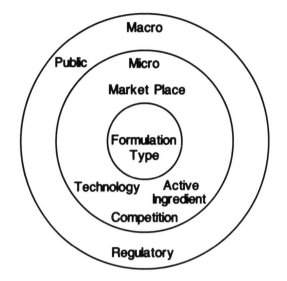

Figure 1. The macroenvironment that influenced pesticidal formulations in the 1960s.

Transition

Three major events occurred to change the nature of the microenvironment. In 1962 the publication of Rachel Carson's *Silent Spring* (*3*) resulted in increased public awareness of persistent pesticides and caused their eventual elimination. During the 1970s a chlorinated dioxin was found as a toxic minor component (*4*) in 2,4,5-T [(2,4,5-trichlorophenoxy)acetic acid]. As a consequence, the detection of minor components is now often reported at the parts-per-million (ppm) and even parts-per-billion (ppb) levels. In 1984, the Bhopal disaster in India was caused by a water–methyl isocyanate interaction. This event led to public concern over chemical plant operations and the environmental safety of chemical storage facilities. The effects are still being felt by the chemical industry (*5*). The impact of these events has been to focus public attention upon pesticides, which in turn has caused a promulgation of governmental, state, and Federal regulations.

Effect of Regulations

In the 1980s regulation became part of the microenvironment that influences pesticide formulations. In Figure 2, the micro-

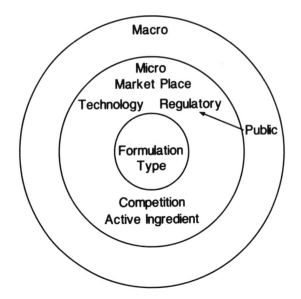

Figure 2. The macro- and microenvironments that influenced pesticidal formulations in the 1980s.

and macroenvironments are shown in which the public influence is strongly linked to the regulatory factor. This relationship is particularly strong in California.

The regulations that have an impact upon formulations are as follows:

Influence	Effect
Safe residues	Set tolerances
Environmental safety	Concern over groundwater issues
Ecological toxicology	Studies on nontarget environment (birds, fish)
Data quality	Development of Good Laboratory Practices
EPA inerts strategy	Categorize inert ingredients
FIFRA 1988	Transfer of pesticides
	Disposal of pesticide containers
SARA 1986	Threshold planning quantities
	Reportable spills
Reregistration	Fewer products

Safe residues in both crops and soils have been defined by setting tolerances. The formulation type can effect residue levels by increasing product longevity with, for example, ultraviolet stabilizers or by using controlled-release technology. Environmental safety relates to groundwater, and ecological toxicology is concerned with the nontarget environment, such as residues found in birds and fish. Again, controlled-release products can modify the rate and therefore the amount of active ingredient leached into the agricultural environment. Avian toxicity, too, can be modified by coating solid granular products to reduce oral toxicity.

Data quality has become a necessary focus for all pesticidal research organizations. The industry works under the guideline of Good Laboratory Practices (GLP), and the EPA section 160.135 is appropriate for formulations (*6*).

Of all the regulations effecting formulation design, the EPA Inerts List published (*7*) in the mid-1980s had the most profound effect. At this time, the EPA has four categories describing inert ingredients:

Category	Definition
1	Inert ingredients of toxicological concern
2	Inert ingredients with a high priority for toxicological testing
3	Inert ingredients that do not fall into lists 1, 2, or 4
4	Inert ingredients with a minimum risk

List 1 contains 57 chemicals, and list 2 contains 61. Of those chemicals in list 2, monochlorobenzene and xylene have the most widespread use in pesticide formulations. However, their use for new formulation registration has been deferred by the EPA until the toxicological database and analysis are complete. In the future, list 2 inert ingredients will gradually disappear, some going to list 1, others to list 3. The outcome will be a diminution in the number of inert ingredients available for formulation. Lists 1 and 2 contain a preponderance of organic solvents; consequently, organic solvent based formulations will decline in usage in the 1990s.

In the near future, The Federal Fungicide, Insecticide, and Rodenticide Act (FIFRA) of 1988 (*8*) will have an impact upon the pesticide industry. The law suggests guidelines for the design and disposal of pesticidal containers with rules that will be proposed on or before December 24, 1991, and will be implemented on or before December 24, 1993. Procedures will describe the safe removal of pesticides from containers and ways to eliminate leaks and splashes. In addition, the act will provide for the safe disposal of containers, their refill, and their reuse. Changes in packaging are already occurring as a result of this legislation. Trends are toward bulk storage facilities. Both recyclable packaging and the development of closed application systems will occur. Some research into biodegradable packaging has made progress, but the logistics of such packaging will be more difficult to implement. Nonetheless, water-soluble poly(vinyl alcohol) (PVA) bags have been used to package WP products for niche markets, and such packaging reduces the user's exposure to potential dusts.

In 1986, SARA, The Superfund Amendment Reauthorization Act, dealt with farm worker exposure and safety issues (9). Title III is a part of the act that is known as the Emergency Planning and Community Right-to-Know Act, which protects farm workers and the public from chemical exposure. It requires that state and local authorities must be notified of the storage of any chemicals that have been declared hazardous and exceed the threshold planning quantities during a year. Thirty products have been listed as likely to be stored on a farm. There are also reportable spill amounts of these chemicals. Implementation of the use of closed application systems should, in the future, minimize the potential for such spills.

The reregistration process for old products has caused a reduction in the number registered for use in the marketplace and thus the number of formulations available.

Another regulatory body, The National Fire Protection Association (NFPA) has issued proposals (10) for suggested implementation in September 1990 for the warehousing of flammable products. Their proposal requires that Classes I and II liquids (flash point <140°F) shall not be stored in plastic containers in general all-purpose warehouses. Three trends are likely to emerge from this proposal: First, plastic containers will decline in use; second, nonflammable formulations will gradually replace flammable products; and third, many warehouses will add fire walls and fire doors.

Effect of Technological Advances

Regulatory influences have not and will not be the only factor in changing end-use formulations. Since the 1960s, the agrochemical industry has changed from an emerging technological industry to a mature one. By 1988, several major products had come "off-patent"; that is, their patents had expired. In order to maintain market share, some companies have developed market segmentation strategies for their maturing product lines by product differentiation. This strategy has been accomplished by product differentiation with novel formulation types or by the introduction of "new" package-mix products with multiple active ingredients, such as Lariat (alachlor–atrazine) and Squadron (pendimethalin–imazaquin). For this to occur, more complex formulation technologies have been developed for both liquid and solid formulations.

Liquid technology employed for multicomponent active ingredient products includes aqueous suspension concentrates (SC), concentrated emulsions (CE), emulsion flowable (EF), and emulsifiable suspension concentrates (ESC). Solid technology has moved toward water-dispersible granules (WDG) and matrix or coated controlled-release products. These technologies are appreciably more difficult to develop.

Concentrated emulsions (CE) consist of an organic phase emulsified in an aqueous phase with the active ingredients in both or either phase. Ordinarily, the droplet size of the organic phase is 0.1 to 1.0 μm, and water is the continuous phase. Maintaining an emulsion for the usual 2-year lifetime of a product is quite a feat. It requires extreme developmental precision to put together such stable emulsions. Careful control of key factors such as pH, ionic strength, phase density, droplet size, and surfactant–emulsifier systems are essential. The American Cyanamid Company has introduced a series of multicomponent package-mix products, such as Squadron, by taking advantage of this technology. Squadron herbicide is a 2.3-lb/gallon CE in which an imazaquin salt is in the aqueous phase and pendimethalin is in the hydrocarbon phase.

Suspension concentrates (SC) contain a solid active ingredient suspended in a continuous aqueous phase. Again, the particle size of the solid is critical and usually is in the 1–10-μm range. Temperature storage stability tests to monitor sedimentation and crystal growth potential are essential in the development process for an SC. Lorox herbicide is an SC product that contains 4 lb/gallon of linuron.

An emulsifiable suspension concentrate (ESC) has the solid active ingredient suspended in an organic solvent. The major difference from an SC is that the continuous phase is now organic. Upon pouring into water, the organic phase of the ESC will form an emulsion prior to end-use application, whereas the aqueous phase of an SC is water miscible. For this reason, the development of an ESC is regarded as a degree more complex than for an SC. The organic phase of the ESC allows many organic soluble pesticides to be formulated. An example of an ESC is a three-component mixture of pendimethalin and alachlor in the organic phase and terbuthylazine as the suspended solid.

Of the four types of formulations, the emulsion flowable (EF or suspensoemulsion) is the most complex. It is a three-phase formulation consisting of a continuous aqueous phase, an emulsified organic phase, and a suspended solid active ingredient phase. Development time for such a formulation is necessarily long. Lariat is an EF in which the alachlor is in the organic and atrazine is the solid phase.

Water-dispersible granules (WDG) have gradually displaced the more dust-producing WP formulations. Essentially, a WDG contains an active ingredient in an agglomerated powder that disperses on water dilution. The solid WDG can minimize chemical degradation and allow for the formulation of chemicals that are either incompatible or unstable in solution. The higher loadings of WDG products, in the range 20–80% active ingredient, afford economies in packaging, transport, and warehousing. Chemical processing of WDG products, is, however, capital intensive, and productivity is often limiting. The Du Pont Company has introduced several of its sulfonylurea herbicides using the WDG technology, for example, chlorimuron ethyl, which is available as Classic. The introduction of WP formulations in poly(vinyl alcohol)-based water-soluble bags has addressed the dusting issue. Such a product is propargite (Omite 30W), an acaricide.

A variety of controlled-release formulations have been explored. Microencapsulation (ME) of alachlor by interfacial polycondensation with hexaethylenediamine and polyethylene phenylisocyanate polymer has led to Lasso:Micro-Tech herbicide. The microencapsulated alachlor is suspended in water as a 4-lb/gallon aqueous flowable product.

Coating of solid active ingredients by nonerodible polymers such as poly(vinyl chloride) (PVC) or ethyl vinyl acetate (EVA) or by erodible coatings such as starch or poly(vinyl alcohol) are being investigated increasingly. Environmental conditions vary so widely that the predicted release characteristics of such products takes considerable testing time to develop. Improved product performance such as a slower release, increased safety to plants or to humans, or reduced water runoff is essential to justify the development of such complex formulations. For this reason, slow-release products are likely to be gradually introduced into niche markets where predictability of efficacy is high or easy to measure. In the 1990s even more novel release systems are likely to be developed. The Landec Corporation is developing (11) thermally activated release pesticidal formulations rather than the erodible barrier products.

On occasion, the active ingredients required for a package mix are chemically or physically incompatible in either solution or solid formulations. The twin-pack technology allows for the rapid introduction to the marketplace of such materials. The twin-pack consists of a container with two separate compartments, one for each active ingredient. These ingredients are brought together only at the point of pouring out from the orifice. Pinnacle is a twin-pack product that consists of isoproturon and imazamethabenz.

The complexity of pesticidal formulations has undoubtedly increased since the early 1960s, yet the time span to develop formulations has diminished. In the 1980s a conservative estimate for the time required for toxicology and registration approval has increased from 4 months in 1980 to about 20 months in 1989. This trend will continue. An implication is that it will take longer to develop and introduce new formulation products in the future.

Veterinary Products

As early as 1953, antibiotics were being marketed in animal feed additives. Some antibiotics were added solely to prevent disease, but the majority were added to feeds for nutritional purposes also, at low subtherapeutic levels in the range 1–100 grams per ton. Examples include chlorotetracycline (CTC), bacitracin, erythromycin, monensin, oxytetracycline, and virginiamycin.

The U.S. Food and Drug Administration (FDA) is responsible for determining if antibiotics in animal feeds are effective and safe for animals and for humans. Any new animal product must have an NADA (New Animal Drug Application) approval. Substantial evidence must be demonstrated of safety and efficacy for its intended use in the target animal and of its safety to humans. An NADA requires manufacturing and analytical methods, control data and assay methods for the detection of drug residues in meat or food. After more than 30 years of use of antibiotics in farm animals, there is no conclusive evidence that their use causes any increase in the incidence or severity of salmonellosis in humans (12).

Administration of an animal health formulation is by either oral, parenteral, or topical applications. The oral delivery can be by mouth, ophthalmic, rectal, or vaginal application. Examples of oral delivery systems include tablet, granules, pastes, gels, powders, solutions, suspensions, capsules, and even bolus forms for slow delivery in the rumen or the reticulum sac of a cow or sheep. Parenteral delivery involves intravenous, intramuscular, intra-articular, or inhalation administration. Examples of parenteral formulations include implants, suppositories, and aerosols. This type of formulation must be sterile, pyrogen free, and, if possible, close to the normal physiological pH. Topical applications rely upon local dermal effect and systemic action. Creams, ointments, pastes, gels, solutions, suspensions, and aerosols may all be used in this manner (13).

A major veterinary product is ivermectin (14); it counteracts parasitic infections. Oral formulations for sheep and pigs, a paste product for horses, and a tablet for dogs illustrate the wide variety of dosage forms used for veterinary products.

The 1990s will see the introduction of a series of biotechnology-derived proteinaceous growth regulants. An example is bovine somatotropin (BST), which increases milk production in dairy cows. Daily oral delivery systems as well as complex controlled-release devices are being developed. Well into the next decade, pulse delivery via an osmotic pump (*15*) or even oral delivery of these proteins may become feasible. In the meantime, the introduction of these growth hormones must gain public acceptance and registration from the Center of Veterinary Medicine (CVM).

Summary

Public, regulatory, competitive, and technological factors influence the formulation of agricultural products. Trends are toward multicomponent active ingredient complex formulations. Nonflammable multiphase liquid products will predominate, water-dispersible granules and coated or matrix-controlled delivery systems will increase in usage, and sustained-delivery devices will be developed for animal recombinant protein products. The conclusion to be drawn from this analysis is that in the 1990s formulations technology will play a pivotal role in the direction of the agrochemical industry.

References

1. Keintz, R. C.; Cross, B.; Kovacs, G. *Pesticide Formulations and Application Systems*; American Society for Testing and Materials: Philadelphia, PA, PCN 08–908000–48, **1986,** *7,* 75.
2. Eto, M. *Organophosphorus Pesticides: Organic and Biological Chemistry*; CRC Press: Cleveland, OH, 1974.
3. Carson, Rachel *Silent Spring*; Houghton Mifflin: Boston, MA, 1962.
4. Kearney, P. C.; Isensee, A. R.; Helling, C. S.; Woolson, E. A.; Plimmer, J. R. In *Chlorodioxins—Origin and Fate*; Blair, E. A., Ed.; Advances in Chemistry Series 120; American Chemical Society: Washington, DC, 1973; p 105.
5. Lepkowski, W. *Chem. Eng. News* **1988,** *66*(1), 8–12.
6. *Federal Register* 54:158, 160, p 135.
7. *Federal Register* 52:77, pp 13,305–13,309.
8. Conner, J. D., Jr.; Ebner, L. S.; Landfair, S. W.; O'Connor, C. A. III "The Federal Insecticide, Fungicide, and Rodenticide Act Amendments of 1988: Summary and Analysis"; McKenna, Conner, and Cuneo: Washington, DC, 1988.
9. Eiche, C. *Indiana Prairie Farmer* **1988,** *160*(5), 1–8.
10. Goerth, C. R. *Packag. Dig.* **1988,** *25*(10), 32–34.

11. Green, C. L.; Stewart, R. F. Presented at the Third Chemical Congress of North America, Toronto, Canada, 1988, AGRO 0159.
12. Fagerberg, D. J.; Quarles, C. L. "Antibiotic Feeding, Antibiotic Resistance, and Alternatives"; American Hoechst Corporation, Animal Health Division: Somerville, NJ, 1979; p 47.
13. Blodinger, J. *Formulation of Veterinary Dosage Forms*; Marcel Dekker: New York, 1983; pp 135–173.
14. Baker, R.; Swain, J. S. *Chem. Br.* **1989,** *25*(7), 692.
15. U.S. Patent 4,036,228, 1977.

Chapter 10

Analytical Chemistry and Pesticide Regulation

James N. Seiber

Profound improvements in analytical methodology have been brought about by the demands of new pesticide regulations. Examples are that analytical detection limits were improved as tolerances were lowered, less stable or more polar pesticide structures were employed, the need for metabolite detection was added, and low-dosage chemicals were introduced. Improvements in analytical turnaround time, accuracy, and precision have also followed changes in regulatory requirements, including establishment of Good Laboratory Practices (GLPs). In some cases, improvements in analytical methodology have stimulated changes in the regulations themselves. Examples include regulatory rethinking of "zero tolerances" as analytical detection limits were lowered, and a series of regulatory steps that were imposed when new analytical methodology showed the presence of toxic residues at low concentration levels, illustrated by 2,3,7,8-tetrachlorodibenzodioxin (TCDD) contaminants in phenoxy herbicides and environmental samples. Thus, improved analytical chemistry is both driven by and a driver of regulatory changes.

The field of trace analysis has blossomed in the past quarter of a century, partly as a result of the introduction of extremely sensitive instruments. Some would argue that our ability to detect and measure trace levels has far outstripped our ability to ascribe biological significance to them. Trace analysis might even be partly to blame for societal confusion over chemical pollution, because the public hears from analytical chemists that chemicals are measurably present and then from agency and industry officials that the levels are too low to be of any health concern. No wonder the public response is to opt for an absolute ban on the chemical pollutants in question, just to be on the safe side. In frustration, analytical chemists are often accused of igniting environmental controversy leading to the loss of useful chemicals and being the "bad guys" who used supersensitive techniques to create headlines. The analytical chemist's rejoinder should be that, if society cannot handle the quantitative data provided by the chemists, someone should try to figure out why and take remedial action rather than blaming the analytical chemist.

In examining the role played by the analytical chemist somewhat more objectively, I will assume that lower detection limits and higher precision and accuracy are desirable so that society will have the most reliable data available for decisionmaking. When required, analytical chemists have responded to the challenges posed by new regulations. But analytical chemists have also spurred regulatory change from such findings as a chlorodioxin or polybrominated biphenyl present in a segment of the environment that should not contain such toxicants. In a broad context, the analytical chemist plays two roles in safeguarding the environment and human health: defining the problem and quantifying the results of actions taken to solve the problem (1). Working in conjunction with toxicologists, who have the expertise to quantify adverse effects in relation to dosage with experimental animals, and epidemiologists, who detect adverse effects among populations by using health-based statistics, the analytical chemist helps to pin down a chemical agent that presents an undesirable exposure situation. In the recent case of the poisoning caused by aldicarb-contaminated watermelons, health officials picked up the makings of an epidemic rather quickly and narrowed the cause of nausea to watermelons; analytical chemists were called in to identify, quantify, and confirm the causative agent (aldicarb and its metabolites). Toxicologists were able to play a supportive role by providing dose–response data that showed that the levels of aldicarb detected by the chemists could in fact cause the observed symptoms. There are many interesting examples of this type of interaction between the three disciplines, a recent one (2) provided by amnesic shellfish poisoning caused by domoic acid (an algal-derived toxicant) in mussels.

Once a "toxics" problem is identified, society can follow various courses of action, ranging from banning or restricting a chemical, to imposing a tolerance or "action-level", to eliminating the source of an effluent, to closing an estuary to public fishing. The analytical chemist may then adopt a monitoring role to follow up the results of the action to ensure that the desired effect has been achieved. Most routine analyses for pesticide residues, carried out by the U.S. Food and Drug Administration (FDA) and other Federal and state agencies, are done for monitoring purposes. In some cases, monitoring data, new toxicology tests, or new epidemiological data will show that the regulatory action did not achieve its intended result, arguing for perhaps more strict regulation or an entirely new course of action.

Advances in Analytical Methodology

Analytical chemists have discharged their responsibility as a primary line of defense against toxic insults by making significant improvements in all steps of analysis. Sampling and extraction of fluid media such as air and water are now frequently done in situ by passing the air or water over an accumulating medium. Solid-phase extraction cartridges, resin cartridges, poly(phenylene oxide) (Tenax) traps, etc., are used routinely in place of solvent extraction, and they provide much lower detection limits by virtue of the larger sample volumes processed. They also eliminate the large solvent volumes, solvent evaporation, and spent solvent wastes associated with the older methodology. Cleanup and resolution are often done with refined or automated chromatographic columns using gel permeation, adsorption, partition, or ion-exchange techniques. But the most visible advances have been made in the determination step, including detection and measurement, most notably with the introduction of highly selective and sensitive gas chromatographic detection.

Table I provides a chronological summary of common de-

Table I. Element-Selective GC Detectors

Detector	Basis for Selectivity	Year First Reported (Approx.)
Electron-capture (EC)	Halogen	1959
Microcoulometric (MC)	Cl, Br, N, S	1961
Alkali-flame ionization (thermionic) (AFID)	P, N	1964
NP-Thermionic selective detector (ND-TSD)	P, N	1974
Electrolytic conductivity		
Coulson (CECD)	Cl, Br, N, S	1965
Hall (HECD)	Cl, Br, N, S	1974
Flame photometric (FPD)	P, S	1966
Thermal energy analyzer (TEA)	NO	1975
Photoionization (PID)	Halogen, S, aromatic	1978
GC–MS (bench top)		
Ion trap detector (ITD)	Diagnostic ions	1983
Mass selective detector (MSD)	Diagnostic ions	1984
Atomic emission detector (AED)	Several elements	1988

tectors, beginning with electron capture (EC) in about 1959. Veteran residue chemists still fondly remember the tritium EC detector (3) mounted in the Wilkins aerograph Hy-Fi model 600.

Tritium EC quickly became a mainstay in analysis of chlorinated hydrocarbon pesticides, and perhaps more than any other technical factor was responsible for the demise of this class of chemicals. Finding accumulations of DDT, dieldrin, aldrin, and their relatives in the body fat of animals and humans demonstrated an undesirable flaw in this class of chemicals resulting from their fat solubility and stability in the environment. This case is perhaps the best documented example of the piecing together of a chemical's environmental behavior profile retrospectively, on the basis of analytical data obtained years after the chemical was first released to the biosphere. There are, of course, many other cases like this.

Other detectors introduced in the 1960s and 1970s were primarily element selective in that their response was to a key heteroatom (halogen, phosphorus, sulfur, or nitrogen) in the classes of analytes of interest (4). Thus, organophosphate insecticides were analyzed by gas chromatography (GC) with a thermionic (alkali-flame ionization) detector or a flame photometric detector; halogenated chemicals were analyzed by microcoulometric or electrolytic conductivity detectors (both of which were less sensitive but more selective than EC); triazine herbicides and carbamate insecticides were analyzed by the nitro-gen–phosphorus thermionic selective detector (NP TSD), etc. The approach was to focus on a unique feature of the analyte (presence of a heteroatom) that could make the analyte stand out from the ocean of natural chemicals present in the matrix that might be observed or interfere with the analyte signal. Thus, background "noise" was reduced and the analyte "signal" was enhanced; the result was an enhanced signal-to-noise (S/N) ratio and thus a decreased lower limit of detection (LOD). Many of the element-selective detectors were developed by pesticide res-idue chemists for application to residue chemistry. The ther-mionic detector, for example, was developed from a chance observation by FDA chemist Laura Giuffrida of an extraordinary enhancement of response to organophosphates after the flame ionization detector became contaminated with salt. In another example, the first reported (5) applications of Hall's electrolytic conductivity detector were to pesticides.

The mass spectrometric (MS) and atomic emission detector (AED) continue this theme of selective enhancement of the S/N ratio. MS zeroes in on diagnostic molecular or fragment ions unique to the analyte structure. The development of the bench-top mass selective detector (MSD) (Hewlett–Packard) and ion trap (Finnegan) were notable recent improvements because they brought GC–MS into the average-sized analytical lab. The atomic emission detector (Hewlett-Packard, 5921A) is the first detector that provides multielement selectivity (heteroatoms plus oxygen and carbon) and, if successful, could largely replace its single-element detector predecessors.

High-pressure liquid chromatography (HPLC) has lagged somewhat behind gas–liquid chromatography (GLC), partly be-cause of the lack of comparable selective and sensitive detection systems. Nevertheless, HPLC can be used for highly efficient cleanup prior to GC (6), for selective determination of com-pounds such as glyphosate using post-column reaction detectors (7), and for confirmation using the emerging technology of HPLC–MS (8). These are in addition to the more straightforward applications of HPLC to the analysis of strongly UV–visible-ab-sorbing or fluorescing analytes that do not require special detectors (7). The future growth of HPLC and its applications to pesticide residues may well occur in the development of supercritical fluid chromatography and microbore capillary HPLC because of the comparative ease with which these techniques can be interfaced with selective detectors.

An unfortunate consequence of the introduction of new, often sophisticated instrumentation is that the expense of conducting trace analyses has increased dramatically. The gravimetric, bioassay, and thin-layer chromatography (TLC)-based methods of the 1940s and 1950s may have been lacking in sensitivity and selectivity, but they were inexpensive and often could be conducted by persons with relatively low skill levels. A typical analysis for a single pesticide in a single matrix, which cost $5–$10 per sample in the 1950s, can exceed $150 per sample in 1990. Not all of the increase is due to instrument, solvent, and technician costs; complying with Good Laboratory Practices and state certification requirements diverts much of the production time to fulfilling paper-work needs, often with little direct relationship to data generation or data quality and a distinctly negative influence on sample throughput.

Immunoassay (IA) offers a welcome diversion from the treadmill of higher costs associated with more sophisticated instrumentation. Immunoassays provide the extreme selectivity of protein–substrate interactions, which excite biologists, with a nonbiological format that is attractive to analytical chemists. Immunoassays involve these steps (9):

1. Extraction of substrate. This step is usually done by using solvent extraction, the same as with conventional analysis. But with some liquid matrices (water, urine, and plasma), it is bypassed completely because IA can be applied directly to aqueous samples without a solvent-based transfer.

2. Cleanup of extract. IA may require some preliminary purification, but usually less than is needed for conventional GC-, HPLC-, or MS-based analyses.

3. Antigen–antibody interaction. An antibody preparation, previously generated in an experimental animal by immunizing it with an analyte-based hapten preparation, is used to react specifically with the free analyte (anti-

gen) in one of a variety of assay formats including ra-
dioimmunoassay, enzyme immunoassays (homogeneous
or heterogeneous), and fluorescence immunoassay.

4. Generation of detectable signal. The presence of analyte
   in the sample is signaled by generation of color, fluorescence, or radiolabel in medium, based on the law of
   mass action.

5. Measurement and interpretation. The data are commonly plotted as percent inhibition versus log substrate
   concentration, an example of which is provided for the
   herbicide paraquat in Figure 1 (*10*). The limit of detection (LOD) is less than 1 ng/mL in this example, which
   is in the range (low picogram to low nanogram) expected
   for most IAs. Just as with GC- and LC-based methods,
   the matrix may affect the LOD, standard curve slope,
   and working range of IAs. This matrix effect and the
   potential for interference caused by cross-reactivity with
   analyte "mimics" in IA need to be carefully assessed so
   that the data are properly interpreted. Figure 2 shows
   the lack of cross-reactivity of various paraquat homo-
   logues, except for the methyl propyl analog, in the paraquat enzyme-linked immunosorbent assay (ELISA). In
   their push to adopt IA methodology because of its much
   higher sample throughput capability, some analysts may
   not have heeded these cautions and thus became disil-
   lusioned with IA.

Some of the barriers to widespread adoption of IA technology in pesticide residue analysis are a rather lengthy method
development time, limitation on the amount of information provided on other substances in the extract, and uncertain regulatory acceptance. Nevertheless, the list of pesticides that have
workable, residue-scale IAs continues to increase, as does public
interest in this relatively new methodology (*11*).

A particularly intriguing development is the use of IA-based
"kits" that can rapidly screen large numbers of samples for specific toxicants in the field or in the laboratory (*12*). This approach is reminiscent of the colorimetry-based "quat kits" used
to screen for paraquat in marijuana samples in the 1970s—an

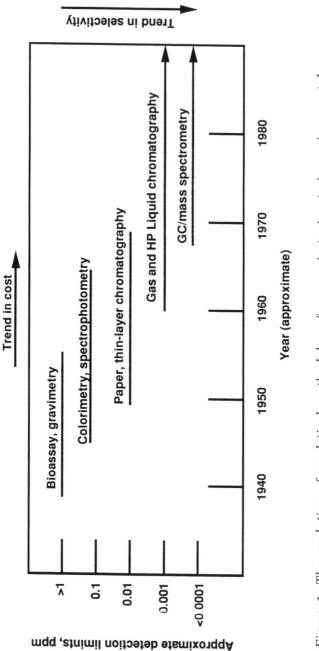

Figure 1. The evolution of analytical methodology for organic toxicants in environmental samples. (Reproduced with permission from ref. 1. Copyright 1982 Plenum.)

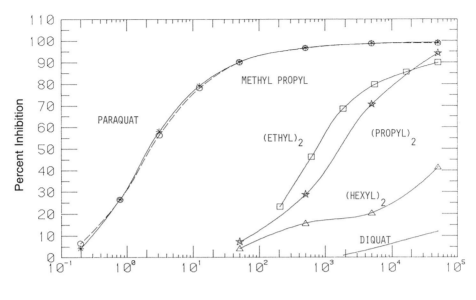

Figure 2. ELISA standard curves for paraquat and five related bipyridilium salts. Results are based on triplicate determinations within run variations of 5% or smaller. (Reproduced from ref. 10. Copyright 1986 American Chemical Society.)

interesting example of the interplay between regulatory policy (U.S. decision to provide paraquat herbicide to neighboring countries to aid in eradicating *Cannabis* plantations), and the entreprenurial abilities of streetwise analytical chemists! The quat kit consisted of sodium dithionite and sodium hydroxide, which were added to water extracts of the material to be tested (dry leaf matter). The blue-colored radical cation of paraquat was the product indicating contamination with the herbicide. The kit fit easily in a coat pocket, and the test took only 2–3 minutes. Cholinesterase inhibition is also used as a basis for screening kits (for organophosphate and carbamate pesticide residues), but IA promises more general utility of the kit concept because IAs can be developed for virtually any analyte class of moderate to high molecular weight.

Demands on Analytical Chemistry Posed by New Regulations

Most regulations that deal with pesticide registration, formulation, use, and inadvertent contamination require something of the analytical chemist. This was true of the Federal Food, Drug, and Cosmetic Act of 1938—the predecessor of the Federal Insecticide, Fungicide, and Rodenticide Act (FIFRA)—through to FIFRA's latest revision in 1988. An early example was the "zero tolerance" concept for potential carcinogens and all pesticides in dairy products. Taken literally, zero tolerance meant that not a single molecule of contaminant was permitted but, practically, it was interpreted on the basis of the concept of no detectable residue using the best available analytical methodology. Zero tolerance was a red flag for analytical chemists, who interpreted it as a challenge to push detection limits to lower levels, so that "zero" residues soon became finite and detectable (13). The outcome was predictable: As detection limits were lowered, tiny residues of carcinogenic pesticides were found; the result was seizures of "contaminated" produce. The best-known case was the seizure of the U.S. cranberry crop just before Thanksgiving in 1959 owing to the detection of residues of a herbicide applied to the crop. (Veteran pesticide residue chemists still refer to the pre-1959 era as B.C. or before cranberries!). Fortunately, finite tolerances were adopted to avoid recurrences of this situation. Even today the zero tolerance concept leads to uneven interpretations between various Federal and state agencies. For example, the FDA and the U.S. Environmental Protection Agency (EPA) will generally ignore a finding of trace residues (<0.01 ppm) of a pesticide on a food crop for which the pesticide has no tolerance, a technical violation of FIFRA. But California's Department of Food and Agriculture (CDFA) has consistently taken a hard-line approach and "red-tagged" or destroyed produce with any concentration level of a nontolerance pesticide. I will return to the likely outcome of the hard-line approach as detection limits continue to decrease, particularly in the mixed cropping patterns that exist in California's agricultural valleys.

One regulatory trend with profound influence on analytical chemistry is the requirement that residue analyses be conducted for all toxic metabolites and formulation impurities, along with

the parent pesticide. Some examples of infamous toxic conversion products include ethylene thiourea (ETU) from the
ethylenebis(dithiocarbamate) (EBDC) fungicides, dialkyl-
nitrosamines from the dialkylamines used to formulate salts
of phenoxy herbicides, unsymmetrical dimethylhydrazine
(UDMH) from daminozide (Alar) growth regulator, and aldicarb
sulfoxide from aldicarb (Chart I). The nitrosamine scare of the
mid-1970s was directly linked to analytical chemistry, in this
case the development of the nitrosamine-specific thermal energy
analyzer (*14*) (TEA) and its coupling as a detector to GC and
HPLC. Detection limits for nitrosamines were lowered dramati-

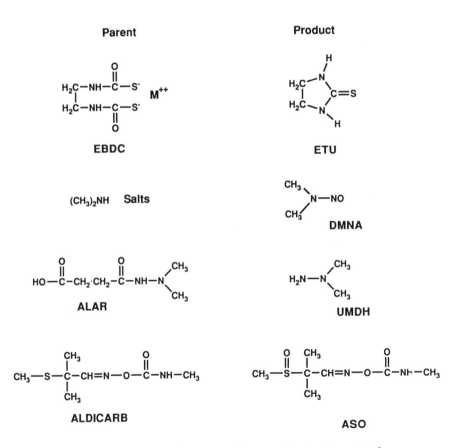

Chart I. Some toxic conversion products of pesticides.

cally, and manufacturers were required to reexamine many pesticide formulations for nitrosamine contamination. The ETU and UDMH stories are still unfolding and have been the subject of several recent reviews, including some in the popular press.

Carbofuran exemplifies the analytical challenge posed by metabolites, as many as five being of regulatory interest (Chart II), including the carbamates 3-keto- and 3-hydroxycarbofuran and the three phenols corresponding to the carbamates. The number of analytical methods and their variants used for carbamate insecticides is unusually large, in part because of the difficulties in chromatographing intact carbamates and also in part because of the need for metabolite analysis within this pesticide class (15). Somewhat surprisingly, EPA continues to accept residue data in which all toxic residues resulting from a pesticide are converted to a single, common derivative. Thus, fenthion (Baytex) residues could be measured as the terminal oxidation product, the sulfone-oxon (16), even though the intermediate metabolites may differ significantly in toxic potency. Although acceptable to EPA, many residue chemists opt for the

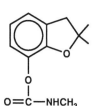

Carbofuran

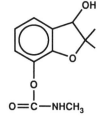

3-Hydroxy

3-Keto

Phenol

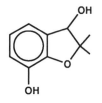

3-Hydroxyphenol

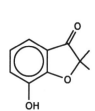

3-Ketophenol

Chart II. Carbofuran and metabolites.

more costly, but toxicologically sound path (*17*) of separately quantifying the parent and each toxic metabolite.

In some cases, the metabolite of a pesticide is the only residue available to assess exposure, and thus must be exploited by the analyst. When EPA and state agencies promulgated regulations requiring the setting of worker reentry intervals for pesticide-treated fields, analytical chemists devised a number of ingenious approaches to assessing external contact (via gauze patches affixed to coveralls and air sampling in the workers' breathing zone) and also absorption of the pesticide in the body through analysis for diagnostic metabolites in the urine. Most organophosphate insecticides—the chemical class of greatest toxicological interest—undergo a combination of oxidation and hydrolysis catalyzed by liver enzymes to produce one or more alkylphosphate metabolites. Because the extracted metabolites are too polar to gas chromatograph as such, they must be derivatized by alkylation to trialkylthiophosphate, trialkylphosphate, or trialkyldithiophosphate esters (*18*). Weisskopf and Seiber (*19*) reported a method in which the metabolites are extracted by using solid-phase extraction (SPE) cartridges; this method replaced a tedious and somewhat distasteful solvent extraction of urine (*20*). The new injector-port trialkylation with tetraalkylammonium hydroxide (*19*) replaced the explosive diazoalkane alkylation (*20*). The new method allows for rapid processing of urine samples and sensitivities sufficient to easily detect both occupational exposure as well as incidental exposures to the parent pesticides.

Analytical Chemistry as a Driver of Regulatory Change

In addition to responding to the needs of changing regulations, analytical chemists have also been prime movers of such change as a result of the development and application of methods that can detect pesticides and their metabolites in environments where they were previously undetected. Examples of chemical targets for these analyses include DDT (and related organochlorine insecticides, as well as polychlorinated biphenyls) in a variety of biotic and abiotic samples, halogenated nematicides such as ethylene dibromide (EDB) and dibromochloropropane (DBCP) primarily in groundwater (well-water) samples, and 2,3,7,8-tetrachlorodibenzodioxin (TCDD).

The finding of chlorinated solvents in New Orleans drinking water in 1975 by electron-capture GC (21) set off a massive search for similar chemicals in fresh-water sources throughout the United States; this search eventually netted several agricultural pesticides. Both EDB and DBCP are nematicides injected into the soil for control of these pests in the root zones of such perennials as grapes, stone fruits, and nuts. In retrospect, the analytical finding of these chemicals in groundwater from the area of use should have been no surprise because of (1) the method of use (injection followed by flood-irrigation of the soils), (2) the application rates (tens to hundreds of pounds per acre), and (3) the physicochemical properties (moderate water solubility and high stability) of the chemicals (22). These properties combine to enhance their migration downward (leaching) to the water table. Both nematicides are now banned, and EPA and several states have enacted groundwater contamination protection regulations designed to prevent future contamination by subjecting chemicals of similar usage and physicochemical properties to special review and use restrictions. The most comprehensive of the state regulations is California's AB 2021 (Connelly Bill), the Pesticide Contamination Prevention Act, which addresses the use of any pesticide whose physicochemical properties have "significant numerical values" lying outside the bounds considered safe by the state on the basis of analyses of prior groundwater contamination records. The use would not be allowed unless the registrant provides proof of no contamination.

The lower limit of detection to which residue methods may be pushed with state-of-the-art methodology can perhaps best be illustrated by TCDD. The extreme toxicity of TCDD to many species, coupled with its environmental stability and food-chain transfer to fatty tissue and milk, has driven the limit of detection (LOD) for TCDD to ever-lower limits. Table II presents this trend, beginning with the use of conventional cleanup and electron-capture GC by Woolson et al. in 1973 (23). A lowering of LOD by 1000× resulted from the introduction of high-resolution mass spectrometry (HRMS) by Baughman and Meselson (24).

Table II. Some Methods for the Trace Analysis of 2,3,7,8-TCDD
in Biological Samples

Samples	Approximate Detection Limit	Techniques	Reference
Bird	50 ppb	Chemical treatment, column chromatography, EC GLC	Woolsen et al. (*23*)
Beef (liver), fish, and crustaceans	20 ppt	Chemical treatment, preparative GLC, HRMS	Baughman and Meselson (*24*)
Trout	10 ppt	Chemical treatment, column chromatography (multiple), HPLC GC–MS	Lamparski and Nestrick (*25*)
Milk	1 ppt	Chemical treatment, column chromatography (multiple), HPLC (*2*), GC–MS	Langhorst and Shadoff (*26*)
Adipose tissue	1 ppt	Column chromatography (multiple—5), GC–HRMS	Patterson et al. (*27*)

Lamparski and Nestrick (*25*) lowered LOD to about 10 ppt by use of a vigorous cleanup sequence prior to GC–MS, which was lowered even more by Langhorst and Shadoff's (*26*) use of an additional HPLC cleanup. Unfortunately, it took two analysts an average of 8 hours per sample to conduct this analysis. Patterson et al. (*27*) achieved the same LOD by using a modified set of cleanup columns and GC–HRMS, but with a substantial improvement in throughput, at least for adipose tissue as substrate.

For many chemicals, a detection limit of 1 ppt may prove impossible using today's methods, and certainly unnecessary for most. But as the TCDD example shows, analysts can work successfully at these levels when health considerations so require.

The Future

The need for more and better analytical data will accelerate developments in analytical chemistry. Improvements in LOD will no longer be the primary goal because today's detection limits will suffice for virtually any situation where concentrations are needed to assess potential health effects. In fact, if LODs are lowered significantly, inadvertent residues will appear in environmental and foodstuff samples, to which they were not intentionally applied, but rather migrated via drift, aerial deposition, runoff, etc. The presence of detectable residues of organophosphate (OP) insecticides on vegetables during the winter season when dormant-spray chemicals are applied to fruit and nut orchards in California's Central Valley may represent the first widespread incidence of "ambient" contamination at levels that are technically illegal but of very doubtful health significance (*28*). In this case, fog water and vapor deposition of the dormant spray OPs may be the source of contamination (*29*). More such situations will undoubtedly be uncovered if routine methods with lower LODs become available. Possibly the nontarget deposition of tiny airborne residues by rainfall, fog, vapor exchange, or dust settling will severely limit the use of pesticides in some areas where little atmospheric ventilation occurs (e.g., in valleys).

The major emphasis will rather be in the development of (1) inexpensive and rapid screening tools for use in the field and also to screen samples prior to full quantitative analysis, and (2) multiresidue methods that can be applied to a wider range of pesticides than accommodated in the existing multiresidue methods of FDA and state agencies. Inexpensive, rapid screening tools include immunoassays and enzyme assays; GC–MS, LC–MS, supercritical fluid chromatography (SFC)–MS, and MS–MS will certainly be examined for multiresidue applications. Miniaturization and automation will also have an impact in both areas as technology advances and as restrictions on the use of laboratory solvents and their disposal costs increase.

For better or for worse, we will certainly have an arsenal of very sensitive analytical methods of ever-improving reliability to meet whatever requirements society presents in its pesticide regulations, and whatever threats arise to human health and the environment through exposures not anticipated by the regulations.

Acknowledgments

The author's work cited here was supported in part by National Institute of Environmental Health Sciences Superfund Grant ES

04699 and by the University of California Toxic Substances Research and Teaching Program Ecotoxicology Program.

References

1. Seiber, J. N. "Analyses of Toxicants in Agricultural Environments", In *Genetic Toxicology*; Fleck, R. A.; Hollaender, A., Eds.; Plenum: New York, 1982; pp 219–234.
2. Quillian, M. A.; Wright, J. L. C. "The Amnesic Shellfish Poisoning Mystery", *Anal. Chem.* **1989,** *61,* 1053A–1060A.
3. Dimick, K. P.; Hartmann, H. "Gas Chromatography for the Analyses of Pesticides Using Aerograph Electron Capture Detector", *Residue Rev.* **1963,** *4,* 150–172.
4. Farwell, S. O.; Gage, D. R.; Kagel, R. A. "Current Status of Prominent Selective Gas Chromatographic Detectors: A Critical Assessment", *J. Chromatogr. Sci.* **1981,** *19,* 258–376.
5. Hall, R. C. "A Highly Sensitive and Selective Microelectrolytic Conductivity Detector for Gas Chromatography", *J. Chromatogr. Sci.* **1974,** *12,* 152–160.
6. Seiber, J. N.; Glotfelty, D. E.; Lucas, A. D.; McChesney, M. M.; Sagebiel, J. C.; Wehner, T. A. "A Multiresidue Method by High Performance Liquid Chromatography-Based Fractionation and Gas Chromatographic Determination", *Arch. Environ. Contamin. Toxicol.* **1990,** *19,* 583–592.
7. Moye, H. A. "High Performance Liquid Chromatographic Analysis of Pesticide Residues", In *Analyses of Pesticide Residues*; Moye, H. A., Ed.; Wiley: New York, 1980; pp 156–197.
8. Voyksner, R. D.; Cairns, T. "Application of Liquid Chromatography–Mass Spectrometry to the Determination of Pesticides", In *Analytical Methods for Pesticides and Plant Growth Regulators*; Scherma, J., Ed.; Academic: New York, 1989; Vol. XVII, pp 119–166.
9. Van Emon, J. E.; Seiber, J. N.; Hammock, B. D. "Immunoassay Techniques for Pesticide Analysis", In *Analytical Methods for Pesticides and Plant Growth Regulators*; Scherma, J., Ed.; Academic: New York, 1989; Vol. XVII, pp 217–263.
10. Van Emon, J. M.; Hammock, B. D.; Seiber, J. N. "Enzyme-Linked Immunosorbent Assay for Paraquat and Its Application to Exposure Analysis", *Anal. Chem.* **1986,** *58,* 1866–1873.
11. Mumma, R. O.; Hunter, K. W., Jr. "Potential of Immunoassays in Monitoring Pesticide Residues in Foods", In *Pesticide Residues in Food: Technologies for Detection*; U.S. Congress, Office of Tech-

nology Assessment, OTA–F–398, U.S. Government Printing Office: Washington, DC, October 1988.

12. "Immunoassays: An Emerging Technology", In *Pesticide Residues in Food: Technologies for Detection*; Herdman, R. C.; Parham, W. E.; Shen, S., Eds.; U.S. Congress, Office of Technology Assessment, OTA–F–398, U.S. Government Printing Office: Washington, DC, October 1988.

13. Zweig, G. "The Vanishing Zero: The Evolution of Pesticide Analyses", In *Essays in Toxicology*; Academic: New York, 1970; Vol. 2.

14. Fine, D. H.; Rounbuhler, D. P. *Environmental N-Nitroso Compounds: Analyses and Formation*; Walker, E. A.; Bogovski, P.; Griciute, L., Eds.; International Agency for Research on Cancer: Lyon, France, 1976.

15. Seiber, J. N. "Carbamate Insecticide Residue Analysis by Gas–Liquid Chromatography", In *Analyses of Pesticide Residues*; Moye, H. A., Ed.; Wiley: New York, 1980; pp 333–378.

16. Anderson, R. J.; Thornton, J. S.; Anderson, C. A.; Katague, D. B. "Determination of Fenthion Residues in Plants and Animals by Electron-Capture Gas Chromatography", *J. Agric. Food Chem.* **1966**, *14*, 619–622.

17. Bowman, M. C.; Beroza, M. "Determination of Fenthion and Five of Its Metabolites in Corn, Grass, and Milk", *J. Agric. Food Chem.* **1968**, *16*, 399–402.

18. "Method for Determination of Metabolites or Hydrolysis Products of Organophosphorus Pesticides in Human Urine, Blood, and Other Tissues", In *Manual of Analytical Methods for the Analysis of Pesticides in Human and Environmental Samples*; EPA 600, 8–80–038, U.S. Environmental Protection Agency: Research Triangle Park, NC, June 1980.

19. Weisskopf, C. P.; Seiber, J. N. "New Approaches to the Analysis of Organophosphate Metabolites in the Urine of Field Workers", Paper presented at the 194th National Meeting, American Chemical Society (AGRO 141), New Orleans, LA, 1987.

20. Shafik, T. M.; Enos, I. T. F. "Determination of Metabolic and Hydrolytic Products of Organophosphorus Pesticide Chemicals in Human Blood and Urine", *J. Agric. Food Chem.* **1969**, *17*, 1186–1189.

21. Jolley, R. "Halogenated Hydrocarbons in New Orleans Drinking Water and Blood Plasma", *Science (Washington, D.C.)* **1975**, *187*, 75–77.

22. "Fumigants and Nematicides under California Conditions", In *Fate of Pesticides in the Environment*; Biggar, J.; Seiber, J. N., Eds.;

Publication 3320, University of California, Division of Agriculture and Natural Resources: Berkeley, CA, 1987.

23. Woolson, E. A.; Ensor, P. D. J.; Reichel, W. L.; Young, A. L. "Dioxin Residues in Lakeland Sand and Bald Eagle Samples", In *Chlorodioxins—Origin and Fate*; Blair, E. H., Ed.; Advances in Chemistry Series 120; American Chemical Society: Washington, DC, 1973; Chapter 12.

24. Baughman, R.; Meselson, M. "An Analytical Method for Detecting TCDD (Dioxin): Levels of TCDD in Samples from Vietnam", *Environ. Health Perspec.* **1973**, *September,* 27–35.

25. Lamparski, L. L.; Nestrick, T. J. "Determination of Tetra-, Hexa-, Hepta-, and Octachlorodibenzo-*p*-dioxin Isomers in Particulate Samples at Parts-per-Trillion Levels", *Anal. Chem.* **1980**, *52,* 1453–1458.

26. Langhorst, M. L.; Shadoff, L. A. "Determination of Parts-per-Trillion Concentrations of Tetra-, Hexa-, and Octachlorodibenzo-*p*-dioxins in Human Milk Samples", *Anal. Chem.* **1980**, *52,* 2037–2044.

27. Patterson, D. G.; Holler, J. S.; Lapeza, C. R., Jr.; Alexander, L. R.; Groce, D. F.; O'Connor, R. C.; Smith, S. J.; Liddle, J. A.; Needham, L. L. "High-Resolution Gas Chromatographic–High Resolution Mass Spectrometric Analyses of Human Adipose Tissue for 2,3,7,8-Tetrachlorodibenzo-*p*-dioxin", *Anal. Chem.* **1986**, *58,* 705–713.

28. Turner, B.; Powell, S.; Miller, N.; Melvin, J. "A Field Study of Fog and Dry Deposition as Sources of Inadvertent Pesticide Residues on Row Crops", Report of the Environmental Hazard Assessment Program, California Department of Food and Agriculture: Sacramento, CA, November 1989; 42 pp.

29. Glotfelty, D. E.; Seiber, J. N.; Liljedahl, L. A. "Pesticides in Fog", *Nature (London)* **1987**, *325,* 602–605.

Chapter 11

Influence of Regulations on the Nature of Newer Agricultural Chemicals

J. W. Kobzina

The growth and development of agrochemical regulations and reregistration requirements have not only had consequences on existing agrochemicals, but they have also profoundly affected newer agrochemicals.

Selection Criteria for Agricultural Chemicals

To understand the influence of regulations on the nature of newer agricultural chemicals, it is first necessary to understand the criteria that are used to evaluate and select pesticides at a given point in time. These selection criteria reflect those considerations that are thought to be relevant and important and are the key factors behind both the evolution and development of newer pesticides and the nature of the regulations that affect them.

The choice of a compound to be developed as an agricultural chemical is influenced by a number of factors, including

1. the nature of the selection criteria that are used to screen candidate compounds, and, by implication, the environmental and toxicological knowledge base that is used to formulate these criteria

2. the availability of a chemical class that can be structur-

2085-9/91/0121$06.00/0 © 1991 American Chemical Society

ally modified to achieve a molecule with the desired properties

3. the nature of the market when the new chemical is introduced

By modification of chemical structures, an organic chemist can both change the physical properties of compounds in a chemical class and alter their biological spectrum of activity. The goal of the structural modifications is a function of the testing criteria that are employed. These selection tests evaluate the biological spectrum of activity, the toxicological properties, and the environmental characteristics of a molecule that is being evaluated.

Before 1970. To appreciate the selection criteria that were used when "older" pesticides were introduced into the marketplace (older meaning before 1970), it is first necessary to understand the nature of the market and the state of our knowledge about the environmental impact of pesticides at that time. Synthetic organic pesticides at the end of World War II were readily received in the marketplace and were looked upon with an anticipation that they would solve many agricultural problems.

For all of the criticism that they later received, the organochlorine insecticides such as chlordane, lindane, and DDT received ready acceptance in the marketplace because they provided very effective control of insects at relatively low application rates and were very cost effective. Their persistence in the environment was viewed as beneficial because it meant that the farmer needed fewer applications in a growing season. Paul Müller's Nobel Prize in 1948 for the discovery of DDT reflected this positive view of the compound's properties.

Generically this older market could be characterized as one in which there were very few agricultural chemicals in the marketplace, many market voids for new chemicals, and a very limited knowledge of the fate of pesticides in the environment. Given this market profile, the historical objective of pesticide synthesis research had at its base one primary question: "Does it work?" In the absence of any significant information about environmental effects of these materials and with a limited ability to analyze for pesticides in the environment, efficacy rather than environmental issues was the chief driving force behind development of these older pesticides.

In this environment, screening processes for agricultural chemicals emphasized three things:

- biological efficacy
- economics
- acute toxicology

Effect of Environmental Concerns. With time there was a growing awareness and understanding of the environmental fate and effects of agricultural chemicals. Many of the properties of older pesticides that had been perceived as favorable attributes were found to have undesirable consequences. The persistence of the chlorinated hydrocarbon insecticides that had been viewed as desirable was found to be a factor in the development of insecticide resistance. This persistence coupled with their lipophilic properties was found to be a primary cause for their bioaccumulation in the food chain. Their broad spectrum toxicity to insects that had been perceived as desirable was found to cross over and be the cause of fish kills when there was runoff into lakes and streams and to affect other life forms such as birds.

The increasing consciousness of the environmental effects of certain agricultural chemicals was coupled with a growing awareness of their presence in the environment as a result of increasing sophistication of analytical instruments. The 1-ppm detection level that was considered state of the art at the time that Rachel Carson's book *Silent Spring* (1) was published has now been replaced by increasingly sophisticated analytical techniques in which sub-part-per-billion detection levels for many pesticides are possible.

This increased body of knowledge served to shape the selection criteria that are now applied to new agricultural chemicals. Whereas historical criteria were based primarily on efficacy, this increasing body of knowledge led to the introduction of significant environmental and toxicological criteria into the selection process.

Reducing the Risk

The increased emphasis on environmental concerns was formalized in the United States with the establishment of the Environmental Protection Agency in 1970 and the passage of the Federal Environmental Pesticide Control Act in 1972. The key to this act was a new statutory standard that reduced the risk of pesticides to humans and the environment. This act affected evaluation protocols for compounds that would be registered and commercialized and also for experimental chemicals.

Application Rate. Prior to any discussion of evaluation protocols for newer agricultural chemicals, it would be useful to look at one factor that is not directly affected by regulations: rate of application. A low rate of application is highly desirable from an environmental standpoint in that lower and lower quantities of chemical are being introduced into the environment. It would, however, be superficial to believe that the increasingly lower rates of newer pesticides are dictated only by regulations and environmental considerations.

With time there has been an increased number of agricultural products in the marketplace. Field performance of newer pesticides is obviously a key criterion for their acceptance. Another measure of their viability would be their cost:efficacy ratio. A good cost:efficacy ratio can be achieved either by the production of fundamentally less expensive chemicals or by a reduction in the use rate. Improved cost:efficacy ratios by means of a reduction in the application rate is a more viable alternative.

Figure 1 plots the year in which a number of key herbicides were introduced versus representative application rates for these compounds. As can be seen from the graph, there has been a gradual but steady decrease in the rate of application of compounds introduced over this time period. Although this reduction in application rates parallels an increase in environmentally driven regulations, it is best viewed as discovery driven rather than environmentally driven. Most of the com-

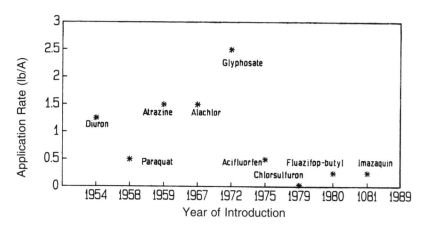

Figure 1. The year of introduction of key herbicides versus representative application rates.

pounds on this graph are representative of distinct chemical classes. Diuron represents the urea herbicides, atrazine represents the triazines, alachlor represents the chloroacetanilides, and so forth.

Reductions in application rates for other members of each of these classes have been quite modest relative to the reductions that have resulted from the discovery of new, more highly active classes of compounds. The new, highly active compounds have been discovered with the help of increased sophistication in plant, insect, and fungal biochemistry; molecular modeling; and a generous dose of serendipity. Because discovery cannot be regulated, the reduction in application rates of newer pesticides is best perceived as driven by the chance of discovery and economic considerations with a coupled benefit of reducing the amount of these chemicals applied in the environment.

One consequence of the emphasis on low application rates has been increasing efforts to separate only the active enantiomers from compounds with a chiral center. With fluazifop-butyl, the elimination of the inactive isomer from the product resulted in a compound that was significantly more active than the mixture of optical isomers and eliminated the inactive isomer from the product.

Environmental and Toxicological Criteria. The characteristics of any new pesticide that is registered must reflect the regulations in effect at the time it is introduced. As newer regulations have increasingly emphasized environmental and toxicological considerations, newer pesticides have reflected these changes. The following is an abbreviated description of environmental criteria that must now be adhered to by newer pesticides.

1. pesticide fate and movement in soil

   - field persistence
   - effect of microorganisms
   - leaching of parent compound and metabolites

2. fish and wildlife safety and hazard evaluation

3. pesticide residue studies, including the effect on nontarget species

4. photodegradation studies

Almost without exception these were not important criteria when the goal was only efficacy.

A similar situation exists for the new toxicological criteria:

1. acute toxicity (the key criterion in historical studies)

   - oral
   - dermal
   - inhalation

2. subacute toxicity

   - oral
   - dermal
   - inhalation
   - effect of residues on nontarget organisms

3. biochemical studies of metabolites

4. reproductive effects

5. teratogenic studies

6. long-term toxicity

7. mutagenic studies

These newer criteria reflect an increased knowledge and emphasis on the possible effects of subacute and chronic exposure to chemicals in the environment.

This first level of understanding of the nature of regulations on the introduction of new agricultural chemicals does not, however, recognize the very deep-seated effect that regulations, particularly as they relate to environmental concerns, have introduced into the screening and selection process for new pesticides. These environmental and toxicological concerns have become part of the culture for the evaluation of new chemicals.

Synthesis and Testing of New Chemicals

The chemical, environmental, and toxicological database that has been built up serves as a guidepost for new chemical synthesis. Given the possibility of work in many areas of chemistry, there is a built-in motivation for a pesticide chemist to pursue areas that do not have environmental problems. If work is pursued in an area of chemistry linked to compounds with environmental problems, screening for these problems will be initiated at an early stage in a compound's evaluation.

Newer synthesis has increasingly emphasized areas that have a biochemical target that does not have an equivalent in humans. Examples of this area in herbicides are the acetolactate synthetic inhibitors such as chlorsulfuron and carotene biosynthesis inhibitors such as fluorchloridone and fluridone. Examples in the more difficult area of insecticides are chitin biosynthesis inhibitors such as diflubenzuron and the juvenile hormone inhibitors.

The growing concern about pesticide runoff and leaching into groundwater has had a significant effect upon testing protocols for new chemicals. Soil properties of a chemical are related to their absorption characteristics on the soil and their persistence in the environment. For chemicals that are soil applied or that are expected to have appreciable soil residues, tests to ascertain their environmental acceptability are conducted at a very early stage. This testing is particularly important for chemicals related to compounds that are known to have environmental problems or that may have undesirable physical characteristics. For these compounds, evaluations to measure their potential for leaching will be conducted in a laboratory by using either soil column chromatography or soil thin layer chromatography. Similarly, persistence can be measured under either greenhouse or field conditions by using both chemical and biological analytical systems. To the extent that problems are discovered on the basis of these early tests, further analogs in a chemical series will be evaluated in an effort to generate a chemical with acceptable soil properties.

Concerns about a chemical's volatility are addressed at an early stage in the evaluation of chemicals in a series when there are laboratory indications that volatilization is occurring, when the compound has structural characteristics related to other compounds that have shown volatility problems, when physical measurements of a compound's vapor pressure indicate that problems may exist, or when there are evidences that a significant loss in the compound's field performance is due to loss of the active ingredient by volatilization. Typical laboratory volatility measurements are performed with a wind tunnel apparatus in which a moving stream of air is passed over a test plot on which the chemical has been applied to either the soil or foliage. This moving air is passed into a detection area in which there is either a chemical or biological evaluation of the amount of chemical that is lost by volatilization.

The nature of very early toxicological evaluations on candidate pesticides has now been broadened considerably because of the increased knowledge base that can now be brought to bear on these compounds. Increasingly, acute, terminal dermal, and dermal sensitization studies are conducted at relatively early stages to confirm that there is no cause for concern relative to applicator safety.

Full-scale animal studies to evaluate a chemical's potential for carcinogenic affects are delayed until well along in a chemical's development, but laboratory screening tests such as the Ames test are frequently conducted at an early stage. This approach is particularly true in the evaluation of novel chemical series or in cases where there is reason to suspect that there may be a problem.

Evaluation criteria have increasingly emphasized environmental and toxicological criteria, and there has been an increasing effort to employ computer programs to predict whether any environmental or toxicological problems would be anticipated. HD's *Topkat* and Case Western Reserve University's *Case* program have found utility in the prediction of toxicological hazards. Compu Drug's *Metabolexpert* has likewise some utility to predict the metabolic fate of chemicals.

In summary, the primary basis for the changes in the nature of newer pesticides resides with our increased knowledge of the interactions of pesticides in and with the environment. This increase in our knowledge base has laid the basis for the various government regulations that cover the registration and use of pesticides. On a parallel path, this increasing knowledge base has incorporated toxicological and environmental considerations into the core of the testing criteria for new agricultural chemicals on an equal footing with the historical, chemical, and biological criteria.

Reference

1. Carson, Rachel *Silent Spring*; Houghton Mifflin: Boston, MA, 1962.

Chapter 12

Biotechnology and New Directions for Agrochemicals

Ralph W. F. Hardy

The agrochemical industry is in a major way a product of biotechnology, the plant biotechnology of the 1930s. This statement is probably a surprise to most agrochemical scientists who view synthetic chemistry and chemotechnology as the generators of the agrochemical industry; DDT fits this view, but 2,4-D does not. Plant biotechnology played a key role with respect to 2,4-D [(2,4-dichlorophenoxy)acetic acid]. In 1924 William Boyce Thompson founded the Boyce Thompson Institute for Plant Research (BTI), a private, not-for-profit research institute initially located in Yonkers, New York (1). Drs. Zimmerman and Hitchcock were hired, and they initiated research on root growth and plant hormones. By the mid-1930s they had discovered that chemicals such as indoleacetic acid, indolepropionic acid, indolebutyric acid, naphthaleneacetic acid, and phenylacetic acid stimulated root growth. Merck marketed Hormodin (indolebutyric acid) as a root stimulant for practical propagation under license by BTI. BTI scientists synthesized 2,4-D in about 1940 and found it to be a most active chemical hormone for root growth (2–4).

These biotechnological studies led to the demonstration of the selective herbicidal activity of 2,4-D and its commercialization as a herbicide by companies such as Du Pont and Dow in the 1940s. Boyce Thompson Institute received modest "royalty" payments. Royalty is in quotations because there was no patent. 2,4-D is still in the marketplace as a herbicide; it has

2085-9/91/0131$06.00/0 © 1991 American Chemical Society

been there for more than 40 years. Annual world-wide sales are still about $250 million, and a conservative estimate of cumulative sales is in excess of $10 billion in 1989 dollars. William Boyce Thompson founded BTI with an endowment of $10 million in 1924. The biotechnological research activities of BTI on root growth led to a product with cumulative sales of $10 billion and a beneficial contribution to world agriculture of possibly $40 billion. Not only did the BTI plant biotechnological studies on root growth and natural plant hormones and their synthetic chemical analogues lead to a major useful product, but these plant biotechnological studies also initiated the herbicide business, which is about 60% of the agrochemical pesticide business. This example documents the impact of an earlier plant biotechnology on agrochemicals.

The impact of the new biotechnology on agrochemicals is all in the future because no products of the new biotechnology for crop agriculture are yet commercial. The most advanced examples are in the early field evaluation stage, and the most significant products may not yet be at even the laboratory stage. Nevertheless, the expected impact of the new biotechnology will be at least as great as that of the plant biotechnology of the 1930s and probably almost an order of magnitude greater because it will change agrochemicals, fertilizers, seeds, food and feed processing, and food and feed itself—and do so in a way that should be environmentally compatible.

Needs of the 1990s and Beyond

The needs of the agribusiness input industry (that is, the suppliers of products for agriculture, such as agrochemicals, seed, and farm equipment), farmers, food and feed processors and distributors, consumers, and society have been changing dramatically in recent years. To illustrate these changes, the needs of two groups, farmers and consumers, will be discussed.

The primary need of farmers of the 1990s and beyond will be increased productivity to maintain or to grow market share in commodity products such as corn, soybeans, and wheat. Productivity is the key to maintaining market share in any nonsubsidized commodity area. Consolidation of some agrochemicals and some fertilizer and seed inputs into only a seed input should improve productivity.

Increased value-in-use food and feed (that is, products valuable for a specific use) will increase farmer profitability in noncommodity areas. In some cases commodities will be converted to increased value-in-use agricultural products. Such higher value-in-use agricultural products may be feeds with higher value as specific-use animal feeds or as consumer foods. The farmer may grow corn tailored to provide a higher quality feed for swine or more healthful oil for consumers.

Crop agriculture in much of the developed world, especially the Corn Belt of the United States, is able to produce more grain than needed within the United States or can be sold abroad. Most of the 1980s has seen huge grain excesses and governmental programs such as the PIK (payment-in-kind) and CRP (Conservation Reserve Program) programs that have sought to decrease grain production. Farmers need new products for nontraditional markets so that the size of the total market for farm products can be enlarged. These new opportunities may include chemicals, polymers, or fuels. Attempts have been made in these areas in the 1980s, but most have been uneconomic, possibly because the characteristics of the crops being used were not tailored to fit the new use but rather were traditional grain crops such as corn, which was forced unsuccessfully to fit new uses.

The agricultural system in use today is less sustainable than is now recognized as desirable. Sustainable agriculture means agricultural practices that can be sustained for a long time. It does not mean elimination of the use of agrochemicals or fertilizers or preservation of the family farm. Sustainable agriculture is profitable and environment-friendly; conserves renewable and nonrenewable resources, germ plasm, and technology; and meets consumer and social needs. Sustainable agriculture will probably emphasize the use of biological systems and deemphasize use of chemicals and costly fossil energy.

A projection of the inherent ability of chemotechnology, engineering, and biotechnology to meet the farmers' needs of the 1990s and beyond is listed in Table I. I make this projection on the basis of my familiarity with these areas and discussions with others. The major conclusion is that biotechnology is inherently more able to meet the future needs of farmers than are chemo-

Table I. Projected Ability of Chemotechnology, Engineering, and Biotechnology Inputs To Meet Farmer Needs

Farmer Needs	Chemotechnology	Engineering	Biotechnology
Increased productivity to maintain–grow market share	+ + +	+	+ + + +
High value-in-use food–feed for increased profitability	+ +	+ +	+ + + + +
New products for nontraditional markets to expand markets	+ +	+ +	+ + + + +
Sustainability–environment-friendly	+ / −	+ / −	+ + + +

KEY: −, negative; +, low, to + + + + +, very high ability rated by author.

technology and engineering. However, chemotechnology is projected to continue to play a significant role, especially in improved productivity.

A similar listing of consumer needs is presented in Table II. The 1989 incidents of contaminated Chilean grapes and daminozide (Alar) in apples illustrate the priority placed on perception of food safety by consumers. Other consumer needs include healthfulness of food (e.g., decreased calories and fat); consumer preference characteristics such as taste, color, flavor, and shelf

Table II. Projected Ability of Chemotechnology, Engineering, and Biotechnology To Meet Consumer Needs

Consumer Needs	Chemotechnology	Engineering	Biotechnology
Safety[a]	−	+	+ + + +
Healthfulness	+ / −	+ + / − −	+ + + + +
Consumer preference characteristics such as taste, color, flavor, and shelf-life	+ +	+ +	+ + + + +
Variety	+ +	+ + +	+ + + + +
Ease of preparation	?	+ + + + +	?
Low cost	+ + +	+ +	+ / −

KEY: −, negative; +, low, to + + + + +, very high ability rated by author.
[a]Based on consumer perception.

life; variety; ease of preparation; and finally, low cost. For the needs other than ease of preparation and low cost, biotechnology is projected as the most relevant technology. Biotechnology may actually increase the cost of food; the improved safety, healthfulness, consumer preference characteristics, and variety should make the consumer willing to pay a premium. Nevertheless, cost of food will probably become increasingly less important in the United States in the 1990s and beyond. The growth of upscale supermarkets and the ability to obtain premium prices for produce that is pesticide-free or organically grown are examples of the growing unimportance of cost of food.

Overall, biotechnology is highly relevant to the needs of consumers and farming. Over the next decade biotechnology is expected to play a major role, and the roles of chemotechnology and engineering will decline but still be significant for several years. Similar conclusions could be made for the needs of agribusiness input industry and food and feed processors.

Agricultural Biotechnology Products and Processes

Agricultural biotechnology products and processes will include diagnostics, microbes and microbial products, seeds, animals, and chemicals. Selected examples are microbes as biopesticides, seeds with added value, and agrochemicals that are designed on the basis of biotechnologically generated information on enzymes and gene function. The most advanced biopesticide and seed products are at the field-test stage, and designed agrochemicals are at the early laboratory stage.

Use of biological agents for control of disease, pest insects, and weeds is of minor importance today. A recent National Academy of Sciences report (5) indicated that biological control should become the predominant means of control. Synthetic chemical pesticides, as well as fertilizers, have been key to increasing food production in the last half of the 20th century for a world population now greater than five billion. However, chemical pesticides and fertilizers may be viewed as "half technologies", and genetically engineered seeds may be viewed as whole technologies, in analogy with Lewis Thomas's earlier description (6) of pharmaceuticals as half technologies and vaccines as whole technologies. Pharmaceuticals and agrochemicals require repetitive manufacture and use for the desired benefits; seeds are self-replicating, and vaccines are used once or infrequently.

Several, but not all, synthetic chemical pesticides have limitations, including persistent residues in soils and groundwater, slow biodegradability, residues in food, decreasing efficiency of discovery and cost (discovery of 2,4-D was much more efficient than is that of the new pesticides), pesticide resistance, unmet pest needs such as viruses and nematodes, inadequate selectivity of beneficial versus pest insects, and decreased industrial competition due to consolidation of agrochemical businesses.

Biotechnology provides the tools to make biological control economically competitive with synthetic chemical control. Biotechnology can be used in many ways. An existing microbial control agent may be made more efficacious, a microbial product may be used as a biopesticide, the gene(s) for a microbial biopesticide may be placed in a more useful host such as another microbe that is more competitive or even placed in the seed so that the plant becomes self-protecting, the cost of production could be reduced, or the delivery system could be improved.

For insect control, biotechnology has been most applied to the bacterial toxic protein produced by *Bacillus thuringiensis*. Genes for *B. thuringiensis* toxins have been characterized, moved to other organisms including crop plants, and are being modified to alter their pesticidal properties. Biotechnology-derived *B. thuringiensis* products are expected to be some of the early commercial ones. Other genes with biopesticidal properties need to be identified and developed so that we do not become overly dependent on a single toxin family. Among the reasons to develop several unrelated genes for pest control is the need to control multiple pests and to avoid development of pest resistance to the biological control agent. Industry already has formed a group to follow and minimize possible pesticide-resistance problems associated with expanded use of the *B. thuringiensis* toxin genes.

Baculoviruses. An exciting potential biological control agent is a large class of insect viruses called baculoviruses. There are more than 500 known baculoviruses, and they are pathogens of many important pest insects, including the gypsy moth and the *Heliothis* and *Spodoptera* complexes. The molecular biology and genetics of a few are described in the *1991 Annual Review of Microbiology* (*7*). Although one or more of the naturally occurring baculoviruses have been registered for commercial use, their limitations as pesticides have prevented their extensive use in the United States. These needs include practical formulation and delivery systems, low-cost process for manufacture, enhanced activity, and greater rate of pest kill.

Scientists in the Plant Protection Program at BTI are applying the tools of biotechnology and bioengineering to develop baculoviruses that will be commercially useful for pest control. In 1989 a baculovirus disabled by genetic engineering was introduced into the field by Wood (*8*). This is the first open-field introduction of a genetically engineered virus. A prior release in Great Britain had used a screened area for containment. The objective of the 1989 field trial was to demonstrate a viral delivery system with a controlled longevity. The longevity in the field was as predicted by laboratory and greenhouse experiments.

Next, genes will be introduced to enhance the pest-insect killing activity of the disabled baculovirus. A possible gene is that for the viral-enhancing protein recently discovered by Granados at Boyce Thompson Institute. This new protein increases viral kill of pests by 25–100 fold. The viral-enhancing factor destroys the periotrophic membrane that lines the insect gut. Other possible genes for increasing insect kill by baculoviruses include those for insect hormones or other toxins.

Economic manufacture of genetically engineered baculoviruses will be necessary for their commercial use. Research at BTI and Cornell is developing bioreactors for virus production. This baculovirus example suggests how biotechnology and associated research may develop efficacious and competitive biopesticides.

Self-Protecting Seeds. Seeds are being made self-protecting by biotechnology. For example, the toxin for *B. thuringiensis* has been incorporated in several plants where protection against pests has been shown in the field. Further genetic modifications may be needed to achieve a commercial level of pest control. Several groups are working in this area. Plants have also been engineered to resist viral damage. Genes for the viral-coat protein provide this protection. Plants with herbicide tolerance may lead to use of less persistent herbicides and give farmers more flexibility in crop protection and crop rotation.

Work on seeds that produce plants with added value is less advanced than that for protection against pests. In 1989, Bio-Technica International field-tested a tobacco with enhanced lysine content. The tobacco is a model for enhanced lysine in crops such as corn so as to make the product of greater nutritional value as a feed.

The tools for production of transgenic plants are becoming routine for several plants such as tobacco, soybean, alfalfa, canola, cotton, and potato, but until recently have not been available for the major grain crops such as corn and wheat. In early 1990, BioTechnica International announced the first transformation of corn with production of fertile seed, indicating that the major barriers to the application of biotechnology to corn had been crossed. The development of transgenic plants with added value will be rapid in the 1990s.

"Ideal" Herbicides. Biotechnology might be used to develop agrochemicals such as herbicides with "ideal" properties. An "ideal" herbicide would have appropriate persistence in the field, effective control of a broad spectrum of weeds, absence of effects on domesticated crops, low manufacturing costs, no human and animal health effects, and potential for proprietariness (that is, ownership of intellectual properties such as patents). The absence of adequate crop selectivity in the past eliminated consideration of candidate herbicides with all of the other ideal properties. Now the ability of biotechnology to make crop plants tolerant to most selective herbicides provides the opportunity to develop those herbicide candidates that were discarded because of inadequate selectivity. This approach could provide some of the most desirable herbicides for the late 1990s and beyond. In all probability, candidates as ideal herbicides have been synthesized and discarded because of inadequate selectivity. This biotechnological approach may provide an opportunity for agrochemical companies with small herbicide market share to regain substantial market share. The approach reverses the traditional approach to herbicide discovery by eliminating the dominant initial requirement for crop tolerance in the screening process.

Other Approaches. Other biotechnological approaches to agrochemical discovery exist. Biotechnology can be used to identify the site of action of an agrochemical. Suborganismal tests using the site of action may assist scoping of active structures without the confusion introduced by the processes of absorption, transport, and metabolism in the whole organism. Directed synthesis of potential agrochemicals may be guided by knowledge of the structure of the active site of the protein target. The structure of the acetolactate synthase enzyme may suggest herbicidal structures other than Du Pont's sulfonylurea and American Cyanamid's imidazolinones. Agrochemicals may be designed to be regulators to "turn on" or "turn off" genes at desired times. For example, such an agrochemical may be used to "turn on" pesticidal genes when pest insect populations become an economic problem. Biotechnology may even provide novel approaches to the problem of pesticide resistance (*9*). These examples of biopesticides, higher value-in-use seeds, and designed agrochemicals illustrate the broad potential of biotechnology to affect agrochemicals.

Proprietariness and Regulation

Biotechnology now enables biological products to be described in molecular terms, as are agrochemicals. In addition, biotechnology is expected to produce biological products or processes with much increased value-in-use. The laboratory phase of biotechnology research is much more costly than is traditional plant breeding. To encourage continued investment in agricultural biotechnology, it is important that there be adequate proprietariness to enable an economic return consistent with the value added and the risk taken.

The trend in the United States and to some extent internationally is increasingly favorable for proprietariness. Hybrid crops such as corn are inherently proprietary. In time, lethal genes that limit the function of introduced genes to a single crop cycle may provide inherent proprietariness in living organisms. During the past decade, the United States Patent and Trademark Office has expanded its protection to include living microorganisms, higher plants, and animals. In the mid-1980s the plant biotechnology group at Molecular Genetics, which was acquired by BioTechnica International in 1989, obtained the first seed patent. It was for a corn that produced additional tryptophan. Proprietariness will enable a return based on the value provided or added by seed and microbe products. This return will apply to all steps in the agricultural food–feed chain. This increase in proprietariness is a most important step in agricultural biotechnology and will, to some extent, make the seed business more like the agrochemical business.

Regulation is also an area of change. In general, products and processes are increasingly regulated. The considerations include environmental and health effects and may expand to include efficacy and possibly socioeconomic effects, as is being proposed by some European groups. Biotechnology products and processes are regulated in the United States by the Environmental Protection Agency (EPA) and the U.S. Department of Agriculture (USDA) for field introduction for testing and commercial use and by the Food and Drug Administration (FDA), EPA, and USDA for food and feed uses. About 100 microorganisms and plants genetically modified by molecular biotechnological processes have been introduced into the field for research in the United States, Europe, and Canada in the past few years. They have been reviewed thoroughly on a case-by-case basis. No surprises have occurred, and it is no surprise that there have been no surprises.

The track record of field introduction of microorganisms and plants modified by traditional processes is highly favorable. Of the hundreds of millions of new plant genotypes developed and field tested by traditional plant breeding in the United States, there are no examples of significant environmental problems. This result is in marked contrast to agrochemicals. The molecular biotechnological processes are more precise than the traditional processes such as plant breeding (Table III) (10).

Recent reports summarize the track record of traditional processes and the knowledge base for molecular processes. A

Table III. Comparison of Organismal, Cellular, and Molecular Processes for Genetic Modification

Parameter	Organismal	Cellular	Molecular
Processes	Breeding	Culture Cell Anther Embryo	rDNA
	Selection Mutation	Regeneration Fusion	
Control over changes	Random	Semi-random	Directed, precise
Primary changes	Unknown	Semi-known	Known
Other changes	Unknown	Unknown	Known
Number of variants needed	Large	Intermediate	Small, in vitro selection methods
Species restriction	Mainly within	Within and across	Within and across
Familiarity	Very high	Intermediate	Low but expanding
Ability to ask and answer risk questions	Low	Intermediate	High
Containment	Dependent on organism and independent of method; established procedures for domesticated organisms		

NOTE: Traditional methods include both organismal and cellular; genetic engineering methods include both cellular and molecular.
SOURCE: Modified from reference 10.

framework for decision-making has been proposed (*11–13*). Current regulation is excessively costly and has precluded field tests for the most part by not-for-profit organizations. Decentralization of the review process should be the next step, so that the thorough, in-depth reviews are focused on those examples where our experience base is limited, or our ability to contain is inadequate, or the risk is greater than negligible.

Structure of the Biotechnology Agribusiness Input Food Complex

The agribusiness input sector is undergoing significant change. Fertilizers have been commodities for several decades. The agrochemical industry is well advanced in a consolidation phase that may leave only 5 to 10 companies existing by the year 2000. Market share for agrochemicals will decrease by 2000 when biotechnological crop protection products become significant.

Seed-company acquisitions by nonseed companies are in the early stage. These acquisitions are being made by agrochemical (ICI, Ciba-Geigy), energy (Shell, BP), and other (Orson) companies. The acquisition strategy is being driven by the anticipation of high value-in-use seeds and seed treatments and the expectation that seeds and seed treatments will be a major growth market. Proprietariness for nonhybrids may make them as profitable as hybrids. The future seed and seed treatments may provide the inputs previously contained in the seed, some of the fertilizers, and some of the agrochemicals.

The focus of the seed and seed treatment business will be on the crop, in contrast to the agrochemical focus that is mainly on the pest(s). The seed and the seed treatment will contain all the genetic inputs necessary for crop protection as well as for value-in-use of the harvested crop. Companies will be identified with specific crops rather than with specific pesticide areas.

Several large established companies, as well as a few development-stage biotechnology companies, are developing products and processes for the seed and seed treatment market. The major value of these seeds may be in their increased value-in-use rather than in their built-in ability to protect against pests. This high value-in-use of the harvested crop may attract relationships with food and feed processors and distributors. Vertical integrations may occur between agribusiness input companies who are developing the seed and seed treatment products, the grower, and the food or feed company with markets for the increased value products. The structuring of such relationships will provide opportunities for creative thinking. The largest companies in the 2000s may be those vertically and horizontally integrated agribusiness food–feed companies rather than the energy, transportation, or information handling companies of the 1980s.

Summary

Molecular biology represents a new era from which the application of biotechnology to crop agriculture will yield biological products or processes for crop protection and for higher value-in-use harvested crops. The products or processes will be diagnostics, microbes, microbial products, seeds, and chemicals. These products will be proprietary and will be highly regulated. Products generated by molecular genetic processes are probably inherently less risky than those generated by traditional genetic processes because of the ability to ask and answer risk questions with greater precision for the molecular than organismal processes. However, the safety record for traditional organismal processes is highly favorable. Established and new agribusiness companies are developing seed and seed-treatment products as well as diagnostics, chemicals, and biopesticides. Vertical relationships between these agribusiness input companies and food-feed processors and distributors may develop. There are inherent major differences between the agrochemical industry and the seed–seed-treatment business.

References

1. McCallan, S. E. A. *A Personalized History of Boyce Thompson Institute*; Boyce Thompson Institute for Plant Research: Yonkers, NY, 1975; 238 pp.
2. *Annual Report*; Boyce Thompson Institute for Plant Research: Yonkers, NY, 1937; pp 6–10.
3. *Annual Report*; Boyce Thompson Institute for Plant Research: Yonkers, NY, 1938; pp 6–10.
4. Zimmerman, P. W.; Hitchcock, A. E. "Substituted Phenoxy and Benzoic Acid Growth Substances and the Relation of Structure to Physiological Activity", *Contrib. Boyce Thompson Inst.* **1942**, *12*, 321–343.
5. National Academy of Sciences, Committee on Science, Engineering, and Public Policy. *Report of the Research Briefing Panel on Biological Control in Managed Ecosystems*; National Academy Press: Washington, DC, 1987.
6. Thomas, L. Remarks given in testimony at U.S. Senate.
7. Wood, H. A.; Granados, R. R. "Genetically Engineered Baculoviruses as Agents for Pest Control", *Annu. Rev. Microbiol.* **1991**, *45*, 69–87.
8. Wood, H. A. "Development of Genetically Enhanced Baculovirus Pesticides", In *Biotechnology for Biological Control of Pests and Vectors*; Maramorosch, K. M., Ed.; CRC Press: Boca Raton, FL, in press, 1991.
9. Hardy, R. W. F. "Biotechnology in Pesticide Resistance Development", In *Pesticide Resistance: Strategies and Tactics for Manage-

Chapter 13

The Fate of Pesticides, the Reregistration Process, and the Increasing Public Concern about Exposure

Keith T. Maddy

Before 1960, the development of pesticides was almost entirely in the hands of entomologists, microbiologists, chemists, and agronomists. Toxicology testing was primarily to determine possible acute effects, and the test results affected label wording only with regard to the possible hazard of handling the formulated products. Relatively few data were available on chronic effects toxicology or on residues that consumers might ingest in foods from crops that had been treated.

In the 1960 to 1980 period, an increasing amount of chronic effects testing that was conducted followed an evolving pattern of protocols. Simultaneously, crop residue testing was extensively conducted, and, when combined with the chronic toxicology data, resulted in the setting of allowable daily intake amounts, acceptable maximum residue levels, and recommended preharvest intervals.

In 1972, the U.S. Environmental Protection Agency (EPA) was created with a mandate to upgrade toxicology testing and to undertake a reregistration process. Even with only the "older" types of data generally available ("older" meaning data collected before 1970), between 1972 and 1984, the U.S. registrations of a number of pesticides were restricted or withdrawn, and in some cases, canceled or suspended. These included DDT, DDD, endrin, lindane (BHC), schradan (OMPA), mobam, terpene polychlorinates mixture (Strobane), mirex, tetrachlorethylene, dibromochloropropane (DBCP), and many arsenical and mercury compounds.

By 1984 the EPA had published in final form entirely new detailed standardized toxicology test protocols (1); these had been developed between 1973 and 1978 and had been used as guidelines for studies started after 1978. The reregistration process now appeared to require that almost all pesticide active ingredients would have to have an entirely new toxicology database developed, except for some acute effects studies. Several registrants submitted a considerable amount of their older data as they began the reregistration process, contending that their earlier studies had produced results that were equivalent to what would be found if they were repeated with the new protocols. In a number of instances the EPA agreed. However, most registrants took no independent action to begin retesting. EPA appeared to many to be moving too slowly in calling in new toxicology test data.

Citizen groups' actions in California in 1985 led to the passage of several environmental laws including Senate Bill 950 (2) requiring California registrants of all pesticide active ingredients to present animal test data meeting the new EPA guidelines on a specific California timetable. Specific reference was made by the law to tests to detect the potential for carcinogenicity, mutagenicity, teratogenicity and other developmental effects, reproductive effects, delayed-onset neurotoxicity, and general chronic effects.

Currently the California review of these new test data and the identification of possible adverse effects is ahead in time sequence of the EPA review process. This situation may change, however, because of a U.S. law change in 1988 by Congress requiring EPA to accelerate its reregistration process (3).

Current Actions

The new 1984 EPA test procedures result in more per-animal chemical exposure to some groups of animals, and the review standards for identifying possible adverse effects are more exacting. Numerous pesticide active ingredients that have adverse effects of possible concern have been identified by California in data that meet the new test guidelines. These active ingredients have been evaluated and regulatory decisions have been made, or they are currently undergoing risk assessments, or they will undergo review as soon as staff time is available. These reviews are primarily of animal toxicology studies and data on human exposure levels, and consequently reveal the risks to all persons who may encounter these pesticides, ranging from applicators to consumers. Many additional new data are still due to be received on most active ingredients being used in California.

The prospect of high costs of data development to meet the newer U.S. and California standards and the likelihood that the presentation and review of such data would lead to severe restrictions or cancellation have been the primary reasons for the withdrawal of many products from California registration in recent years. Some of these registrations had been inactive. These withdrawals have resulted in the complete removal of 243 pesticide active ingredients from use in California; a high of 888 in use in 1985 was reduced to 645 by mid-1990. Of these 243 withdrawn registrations, 27 were from a California list of 200 active ingredients of highest priority of concern. During this time period, only 20 new active ingredients were registered in California; 22 others were in the preregistration process in California in mid-1990, which is now often a several-year review process for new active ingredients. A sizable number of pesticide active ingredients still registered are disinfectants and sanitizers and not the chemicals used as insecticides, fungicides, and herbicides the public so often only thinks of as pesticides.

Animal tests yield three types of test data of particular concern to the general public: cancer, birth defects and other developmental effects, and adverse male and female reproductive effects. Animal tests required by the U.S. Food and Drug Administration (FDA) for drugs are very similar to those EPA requires for pesticides. Controversial evaluation procedures for cancer-producing potential are currently being used to determine if the extent of tumor development is of statistical or biological significance. I have observed that, with tests conducted for both EPA and FDA for cancer, for more than half the chemicals (drugs and pesticides) tested, an excess of tumors is found in at least one sex of one species. Also, with tests for general reproductive births in the United States ranges from 4 to 7%. This incidence has not increased or decreased significantly overall during the past 40 years (4). However, some persons and public interest groups are concerned that several of the large numbers of organic chemicals that have been synthesized since the 1950s and formulated into drugs, pesticides, and other household products have increased the incidence of birth defects in exposed women. This belief can probably be traced largely to the thalidomide disaster in Europe in the 1960s, when use of this chemical as an anti-emetic by pregnant women was found to cause teratogenic damage to the developing limbs of thousands of children before they were born.

One of the problems in selecting suitable test animals to test for teratogenic potential has been the wide variation in dosages that cause teratogenic effects in various species of mammals. For instance, the rabbit requires 25 times and the rat requires 50 times the dosage of thalidomide that produces birth defects in humans. Other than thalidomide, few agents have been proven to cause human birth defects; those that are believed to cause human birth defects include the German measles virus, the organism causing toxoplasmosis, the organism that causes syphilis, radiation, Warfarin, androgenic steroids, morphine, aminopterin, and dimethylmercury.

On the other hand, the currently used studies in pregnant animals have shown many substances to be animal teratogens. Toxicologists do not agree on what these animal test results mean in relation to human exposure. Some of these may some day prove to be human teratogens. Substances found to be positive in such animal tests include vitamin A, nicotinic acid, insulin, estrogens, epinephrine, penicillin, tetracycline, streptomycin, boric acid, sulfonamides, thallium, selenium, various dyes, nicotine, caffeine, and aspirin. Aspirin is often used as the positive control in animal teratogen studies. It produces birth defects in animals at low dosages, yet epidemiological studies of pregnant women who have ingested sizable quantities of aspirin during their pregnancies have not shown an excess of birth defects. In some pregnant women, however, high doses of aspirin have caused bleeding from the uterus.

For caffeine, on the basis of some animal test data, a fetus would have only a 20-fold safety factor after the mother's consumption of one cup of coffee. A more desirable safety factor conventionally would be at least 100-fold.

In spite of these uncertainties, as well as many uncertainties on how to interpret the animal toxicology studies, for pesticides the EPA now requires these types of teratogenic studies on two species of animals (usually the rat and the rabbit) if there is any significant potential of human exposure. If the test in either species is positive, the EPA requires a large safety margin to avoid excess exposure of humans, or they phase out the use of the pesticide entirely on the basis of the animal studies, even if actual human risks cannot be proven.

Adverse Reproductive Effects to the Male and Female. Adverse effects of various chemicals found in reproductive effects studies in test animals are of increasing concern. In the newer studies on cholinesterase inhibitors, some of these adverse effects are found to begin at the point that cholinesterase levels show depression. These results raise policy questions about past guidelines that allowed levels of up to 50% of plasma and 40% of red cell cholinesterase depression for exposed workers (including females who might be pregnant).

Female reproductive toxicity can be defined to include effects on the adult or, where appropriate, developing female organism, including, but not limited to

1. adverse effects on reproductive structure or function, such as

   - genetic damage to the ovum or its precursors
   - alterations in ovulation or menstrual (estrous) cycle, or menstrual (estrous) disorders
   - impaired or altered endocrine function
   - complication of pregnancy

2. impaired reproductive performance (e.g., subfertility or infertility), including

   - increased pregnancy wastage (e.g., miscarriage or stillbirth)
   - inability or decreased ability to conceive (e.g., increased time to conception)

- adverse effects observed in sexual behavior, onset of puberty, fertility, gestation, parturition, lactation, or premature reproductive senescence

Male reproductive toxicity is defined to include effects on the adult or, where appropriate, developing male organism, including, but not limited to

1. adverse effects on reproductive structure or function, including

   - genetic damage to the spermatozoon or its precursors
   - impaired sperm or seminal fluid production, including alterations in sperm number, morphology, motility, and ability to fertilize
   - impaired or altered endocrine function

2. impaired reproductive performance (e.g., subfertility, infertility, or impotence)

Cancer

Another example of regulatory uncertainty involves interpretation of animal cancer test data (5). At least 30 chemicals or processes are widely accepted as proven human carcinogens. Of these, only inorganic arsenicals and chlorine added to drinking water are also used in pesticides. On the other hand, the current chemical testing regimen in animals in so rigorous and involves such high dosage levels of the test chemical that many of the chemicals put through the regimen produce some excess of tumors in at least one of the exposed groups of animals. Part of this finding may result from an attempt to save money by combining chronic toxicity studies and cancer studies for the same chemical. The high-dose group of the combined study may experience such extensive chemical overexposure and disruption of metabolism with so much cellular damage that some excess tumors may occur.

Some toxicologists believe this high-dose group examined for chronic effects should often be ignored in making cancer risk extrapolations. Others are concerned that the mouse, the test animal in which positive results most commonly occur, may not be an appropriate test animal to predict human carcinogenicity. There are also concerns when only liver tumors are found because some toxicologists believe that this effect is a poor predictor of human carcinogenicity. Also, positive results in a gavage-dosing study may indicate false-positive results as far as humans are concerned if human exposure is only by inhalation, through the skin, or by less intensive dietary exposure.

Drugs and pesticides are considered for approval on the basis of these animal test results, and there is ongoing controversy about how to interpret these studies. However, where knowledge is lacking, regulatory agencies tend to the conservative side.

The U.S. Supreme Court's plurality decision on workplace exposure to benzene stated that workplace risks from exposure to a carcinogen with a cancer risk rate not exceeding one chance in 1 billion would be acceptable. On the other hand, it stated that the "reasonable" person would attempt to reduce this risk if it were worse than one chance in 1000. This level of risk was initially used as the guideline in protecting industrial workers in the United States. A number of nonpesticide chemical exposure situations still exist where lifetime cancer risks are greater than one chance in 1000 when using the current available test data. This risk level assumes that the animal cancer test results are in fact directly predictive of human cancer–causation potential.

When animal test data suggest that a very useful pesticide is a carcinogen, the California Department of Food and Agriculture attempts to maintain registration if calculated additional risks are or can be reduced to negligible levels. In such cases the government attempts to ensure that consumers face a calculated lifetime additional cancer risk of no more than one chance in 1 million. For lifetime exposures of applicators and of field workers reentering treated fields, additional risks of one increased chance in 100,000 are usually considered acceptable.

In regard to consumer exposure, the major carcinogenic risk from exposure to pesticides involves the use of chlorine in drinking water. This exposure involves several million Californians. The chlorines combine with biologic material and form chloroformlike chemicals. Some lifetime risk calculations place the risk as great as one increased chance in 10,000 of contracting cancer from this source. Of course, the alternative of not using chlorine and acquiring very serious and sometimes fatal enteric diseases is unacceptable.

There is considerable controversy over the statistics used to convert animal cancer data to possible human cancer risk levels, as already discussed. The currently used statistical procedures probably calculate risks far in excess of the real human risk because of the great biological uncertainty in extrapolating from high-dose test animal results to very low dose human exposure. More important is the method of low-dose extrapolation.

These calculated risks can be compared with the actual current human carcinogenic risk. With the exception of cancer associated with cigarette smoking, the overall human cancer incidence rates have not changed significantly in the United States in the past 50 years. Some persons and public interest groups had been concerned that the great increases in the use of organic chemicals since 1950 might cause a significant increase in cancer in the human population.

The current lifetime risk of dying of cancer in the United States is about one chance in four. For a person who smokes heavily, eats large amounts of fatty foods, and consumes considerable amounts of alcoholic beverages, the lifetime cancer risk may approach one chance in one. The current lifetime risk of dying of cancer for the person with good health habits, at best, is one chance in 10. When government regulatory officials compare these current cancer rates with the conservative standards being employed for pesticides used in the workplace and for consumers, the one additional chance in 10,000 up through one chance in 1 million does not appear tremendously hazardous.

All possibilities to improve the ability to determine actual carcinogenic potential for chemicals should be employed. Chemicals should not be used as pesticides in a manner with any significant cancer risk, but regulators must think seriously about what "significant" means, and probably should look elsewhere for the primary causes of human cancer. Bruce Ames, in an article in *Science* (6), argued that many substances that we have accepted as normal food components are significant mutagens or carcinogens. He stated that these problem chemicals naturally occurring in our foods appear at levels of up to 10,000 times higher than levels of regulated pesticide residues. These chemicals include safrole, hydrazines, furocoumarins, solanine, quercetin, quinones, theobromine, pyrrolizidine, allyl isothiocyanate, gossypol, sterculic acid, leguminous plants, sesquiterpene lactones, phorbol esters, and canavanine. These are found at substantial levels in foods such as black pepper, mushrooms, celery, potatoes, coffee, cocoa, figs, parsley, horseradish, and charred foods. Ames' position has been supported by Weisburger (7), Miller and Miller (8), and Lai and Woo (9).

The difficulties in regulating substances found to produce cancer in test animals were recently illustrated by a U.S. National Toxicology Program (NTP) report (10) on carcinogenesis studies on d-limonene in rats and mice. d-Limonene is a natural component of many fruits (especially citrus), vegetables, meats, spices, and other foods. It may serve as a natural pesticide in citrus fruits. It is naturally found in orange juice at levels near 100 ppm. Because of its lemon odor and citrus taste, more than 1 million pounds per year are extracted from citrus peel and pulp for use as a food-flavoring agent, in household fragrance sprays, in cosmetics, and in perfumes. It has been used in the following foods and detected at the levels indicated in parentheses: soft drinks (31 ppm), ice cream (68 ppm), candy (49 ppm), baked goods (120 ppm), gelatins and puddings (48–400 ppm), and chewing gum (2300 ppm). It has been detected in drinking water and in air pollution studies (5.7 ppm in air over Houston, Texas).

d-Limonene is also registered and used as an insecticide to kill fleas on dogs and cats by topical application and for an indoor space spray. It is also a component in the following pesticide active ingredients: oil of citronella, oil of citrus, oil of lemon, and oil of orange.

The NTP report stated that there was clear evidence of carcinogenic activity of d-limonene for male rats as shown by increased incidences of tubular cell hyperplasia, adenomas, and adenosarcomas of the kidneys. Because the treated female rats and the male and female mice did not demonstrate carcinogenic activity, this pesticide active ingredient may get a low "possible" human carcinogen classification by EPA. A decision to phase out all animal carcinogens from use as pesticides (as suggested by some proposed California law changes) might eliminate *d*-limonene as a pesticide, while the major exposure of people from nonpesticide uses of *d*-limonene would continue.

Actual and Perceived Risks

One should also compare the various calculated risks of dying of cancer to other risks people encounter, such as dying in a automobile accident. Currently in the United States, for each citizen there is one chance in 70 of dying in an automobile accident in a lifetime. Stated in another way, in the United States for every 50 miles a person rides in an automobile, that person has one chance in 1 million of being killed in an automobile accident.

In the United States, currently, with the combination of large agricultural crop surpluses subsidized by taxes and a growing consortium of environmental and public health groups interested in clean and pure food, the public is becoming reluctant to accept continuing exposure to pesticides in air, water, or food, or occupational exposure.

The U.S. public is exerting increasing pressure to test more imported food and to ban entry of any foods that contain residues of pesticide active ingredients not now registered in the United States for the same use. This ban includes chemicals previously registered as well as those never registered. FDA is now beginning to test imported foods for 124 pesticide active ingredients used on food crops in other countries but never registered for use in the United States.

Recently in the United States, many members of the general public have indicated that they have limited faith that the government is ensuring the best possible food supply. They believe that food production technology by now should make it possible that the natural untreated food can be provided to the consumer without pesticides, preservatives, hormones, antibiotics, or additives of any type.

It is easy to state that certain public interest groups dominate the media's attention with misstatements and half-truths and that if the public is simply told that natural toxins in our foods are up to 10,000 or more times toxic than most pesticide residues, then the public will accept the current level of chemical residues in our food supply. This expectation is not realistic. A majority of U.S. citizens now seem to be convinced they are being unnecessarily exposed to chemicals that have been added to their foods.

My opinion is that in California and probably also in the rest of the United States, at least half of the pesticides now registered will probably not be registered 10 years from now because of concern about continued use of any active ingredient that produces a significant adverse effect in the animal toxicology studies submitted in the reregistration process, unless exposure can be kept to negligible levels and only near-zero residues can be found in harvested crops. This opinion is based upon the following conditions:

1. the current near-universal lack of a complete up-to-date toxicology and crop residue database on any of the pesticide active ingredients still registered

2. the need to repeat some studies or follow-up studies even if new protocols are used

3. the multimillion dollar cost of such data development for each active ingredient (U.S. chemical companies who in the past synthesized, tested, and developed many chemicals for use as pesticides have for the most part determined that there is decreasing economic benefit in continuing to develop such chemicals. Many have stated that they will develop new data only for a few major-use chemicals that have a chance of profitability and will maintain other registrations only so long as additional testing costs can be avoided. Several are selling or eliminating their pesticide businesses.)

4. the unavailability of enough laboratories to undertake such a volume of testing

5. the low volume of uses of many of the pesticide active ingredients still registered

6. new concerns and testing costs of the "inert" ingredients

7. the high probability of most active ingredients demonstrating one or more toxic effects of such extent that an exposure assessment will be needed

8. the lack of exposure information on each use for most active ingredients

9. the concern about increasing levels of exposure to a pesticide active ingredient as many alternate pesticides are phased out and are no longer available to dilute exposure to a specific active ingredient

Furthermore, the U.S. reregistration process has really just begun, and a myriad of findings of adverse effects in currently registered chemicals yet to be revealed will further erode public confidence. Also, scientific progress is being made in developing several neurobehavioral tests that are likely to be added in the next 5 years to the routine U.S. testing procedures on drugs and pesticides. Some of these tests are quite sensitive and may reveal behavioral effects, especially in the newborn of treated animals, that will raise additional concerns about continued use of these chemicals.

Finally, although the antipesticide movement is not as strong outside the United States, regulatory agencies around the world are spending increasing amounts of time explaining why a particular pesticide is being used in their country when it is not registered in such countries as the United States, Canada, United Kingdom, Germany, Japan, or other countries that have a rigorous health and safety pesticide review process.

Discussion

Developed and developing countries may have to deal with a number of questions in view of this increased concern about pesticide exposure. This problem is growing, particularly for developing countries.

Pesticides Banned Elsewhere. How can a country find out what pesticides have been banned elsewhere or are under consideration for restriction or banning?

1. Subscribe to newsletters such as *The Pesticide* and *Toxic Chemical News*, which is published weekly in Washington DC.

2. Contact the U.N. Food and Agriculture Organization (FAO) or the World Health Organization (WHO).

3. Contact the EPA for current lists of "special reviews" and various restrictive actions.

4. Contact the California Department of Food and Agriculture (CDFA) for a current list of active ingredients under review because of possible adverse effects.

Use of Unregistered Pesticides. For what reasons may a developing country continue to use or start using a pesticide that is not registered in the United States or another country that has a rigorous toxicology review program?

1. The developing country has a pest–crop combination that does not exist in the United States or other developed country.

2. The developing country has a public health problem that dictates use of a moderately hazardous chemical to prevent a significant incidence of human illness.

3. The developing country has a human malnutrition or a starvation problem that far outweighs moderate or high risk hazards from certain pest-control chemicals.

4. The developing country has an epidemic upsurge of a particular pest that threatens a major food source.

5. A ban in the United States or elsewhere may have been a result of possible direct contact by pregnant females before, during, or after application. Use in the country of concern may not involve contact by any females, or contact could easily be prevented. This assumes there is a large safety factor for a pregnant female consumer of treated crops.

When Not To Use Pesticides. When would a developing country usually advocate not using pesticides that are not used in developed countries?

1. The food crops are destined for export to a developed country.

2. Meat, milk, or butter from treated animals or animals fed treated crops is then exported to developed countries.

New Pest-Control Research. What are some examples that new pest-control research should now concentrate on?

1. Improved integrated pest management processes such as waiting until pest threshold levels are exceeded, avoiding large area monocultures of a single field crop, crop rotation, etc. This needs to be the main thrust of pest control education in agricultural colleges worldwide. Far too many staff members educated in the 1950s and 1960s are still extolling the virtues of the chemical revolution through pesticide use as being capable and acceptable for solving almost all pest problems.

2. Organic food production schemes that use little or no synthetic pest control chemicals

3. Pheromones

4. Viruses, bacteria, fungi, and predator insects that control specific pests

5. Growth regulators

6. Antifeeding agents

7. Chemicals with very high insect or mite toxicity and very low mammalian toxicity such as pyrethroids (neurotoxic potential) and avermectin (birth defect causation potential), both with large safety factors because of low dose levels used

8. Substances that the public considers "natural" such as sulfur, lime, chrysanthemum leaves, garlic, limonene, soaps, vegetable oil, waxes, silica, chitin, and perhaps even ground tobacco leaves

9. Chemicals of low toxicity for mammals that will kill pests for a few hours or days and then quickly degrade to substances with negligible toxicity

10. Repellent or attractant chemicals placed near the foliage of food crop but not on it

11. New mechanical procedures such as vacuuming foliage to suck off insects or mites or water jets to wash pests off plants

12. Physical procedures such as heat, cold, or freezing
13. Gamma radiation. Although a vocal segment of the general public misunderstands this technology, it deserves more development for post-harvest treatments.
14. Natural plant breeding for pest resistance
15. Genetic bioengineering to develop pest-resistant plants

Acknowledgments

This chapter was originally presented at the First Asia–Pacific Conference of Entomology, November 8–13, 1989, in Chiang Mai, Thailand.

References

1. "Environmental Protection Agency, Data Requirements for Pesticide Registration", *Federal Register* Oct. 24, 1984; *Code of Federal Regulations* Title 40, Pt. 158.
2. "Pest Control Operations", Article 14, Birth Defects Prevention, Sections 13121 through 13130; Food and Agriculture Code, Div. 6, State of California, 1985.
3. Federal Insecticide, Fungicide, and Rodenticide Act, amendments of 1988, Public Law 100–532, *United States Statutes at Large* 102, p 2654.
4. Fraser, F. C. "Relation of Animal Studies to the Problem in Man", In *Handbook of Teratology*; Wilson, J. G. W.; Fraser, F. C., Eds.; Plenum: New York, 1977; p 75.
5. Furst, A. "Yes But Is It a Human Carcinogen?" *J. Am. Coll. Toxicol.* **1990,** *9,* 1–18.
6. Ames, B. N. "Dietary Carcinogens and Anticarcinogens: Oxygen Radicals and Degenerative Diseases", *Science (Washington, D.C.)* **1983,** *221,* 1256–1264.
7. Weisburger, E. K. "Natural Carcinogenic Products", *Environ. Sci. Technol.* **1979,** *13,* 278–281.
8. Miller, E. C.; Miller. J. A. "Naturally Occurring Chemical Carcinogens That May Be Present in Foods", In *Biochemistry of Nutrition*; Neuberger, A.; Jukes, T. H., Eds.; University Park Press: Baltimore, MD, 1979; Vol. 27, pp 123–165.
9. Lai, D. Y.; Woo, Y. T. "Naturally Occurring Carcinogens: An Over-

view", *Environ. Carcinog. Rev. J. Environ. Sci. Health (Part C)* **1987,** *5,* 121–173.

10. "Toxicology and Carcinogenesis Studies of *d*-Limonene (CAS No. 5989–27–5) in F344/N Rats and B6C3F Mice, Gavage Studies", National Toxicology Program, Technical Report Series No. 347, U.S. Dept of Health, Education, and Welfare, Public Health Services, National Institutes of Health: Bethesda, MD, 1990.

Chapter 14

Agrochemicals in the Future

Gustave K. Kohn

Although no one can make infallible predictions, I intend to look at trends—economic, legislative, and scientific—and extrapolate from them to the future.

Perceptions, Regulations, and Legislation

The first two letters in the May 19, 1989, issue of *Science* (1) discussed the issues relating to Alar (daminozide) on apples. The first was written by an earnest and involved consumer advocate and ardent environmentalist and the second by a distinguished biochemist and his collaborators at the University of California at Berkeley. The first letter states "Risk management must balance values and ethical choices and is unavoidably a political, not a scientific process." This opinion epitomizes the problem facing government regulators, the agrochemical industry, the scientific community, the ACS, and members of the ACS Division of Agrochemicals. (It also leaves the perhaps unintentional implication that our scientific community is devoid of value systems and ethics.) Furthermore, ethical activity and values *require* broad understanding and knowledge, or else we suffer the value and ethical distortions of orthodoxies based either on inflexible interpretation or ignorance, as are currently exhibited by various fundamentalists on social and political issues.

Many concerns over the environmental consequences of modern technology, including agricultural technology, are legitimate. There are two paths before us. One is to halt the technology. The other is to holistically evaluate all aspects of the societal effects from that technology. Following analysis, we search for and execute new technology that reverses or ameliorates the undesirable environmental consequences that resulted from the earlier technology. This struggle is never ending. It is life itself, and it brings contemporary meaning to the eating of the fruit from the tree of knowledge. We must perpetually struggle.

Most likely, the future will bring a continuation of the conflict over the environmental and health aspects of technological advances. Until the advent of the 21st century, much irrationality with regard to these problems will continue. It will be reflected in legislation and regulation that restrict technological advances both in the United States and Europe, to be followed after, at, or near the year 2000 with a return of appreciative understanding of the contributions of science and engineering. The American Chemical Society, the energy industries, the American chemical industry, and particularly the agrochemical industry will experience a period of difficult struggle and increased economic pressure. More restrictive legislation that can be expected during the next decade will pertain to the use of agrochemicals, including the products coming from biotechnology. Ultimately, the ill consequences flowing from these restrictions will serve to educate both the general population and the legislators.

Students and Education

An appalling drop has occurred in student enrollment in the hard sciences, that is, chemistry and physics. The following statistics (2) illustrate the situation for chemistry.

Year	No. of Ph.D Graduates in Chemistry	Total No. Choosing Teaching of Chemistry in College and University
1970	2223	18.5
1985	1836	8.0

It is deplorable that bright, young U.S. students choose business administration and law rather than science. Making money is the prevailing value throughout our affluent society, even among students. The desire for immediate satisfaction is not characteristic of our youth alone. An electorate that chooses to ignore budget imbalance and trade deficits, whose industrial management leadership concentrates on short-term objectives rather than long-term planning provides the example that our youth exhibit in their educational attitudes.

For most agricultural colleges in the United States, an enrollment decline through the 1990s is expected. The figures for 1978 through 1987 support this expectation (*3*).

Year	Enrollment	Year	Enrollment
1978	98,030	1983	80,217
1979	97,177	1984	77,772
1980	92,833	1985	74,885
1981	91,984	1986	68,131
1982	86,842	1987	63,919

The prestigious agricultural colleges are, and for the near future will be, recruiting students from foreign countries. The University of California at Davis College of Agriculture is perhaps unique in that each of the crop and animal husbandry fields has included a molecular biology component. For bright young minds, molecular biology is an attractive and challenging new frontier, just as electronics was in the preceding decades. As a result, in a period of general decline in enrollment of schools of agriculture, the student population there has actually increased and is expected to continue at least at a steady state. The agrochemical industry and government laboratories greatly depend on the students from agriculturally related institutions.

Another change has occurred in the student bodies of these colleges of agriculture. Whereas in the 1970s the gender of the student body was almost exclusively male (women studied home economics), the current enrollment at the University of California at Davis is slightly more than 50% female, and there are increasing numbers of Asian, black, Hispanic, and other minority students. Both industrial and government laboratories will reflect this change in the next decade.

In addition to these predictions, there will be a continued need for analytical chemists, an increased demand for biochemists and enzymologists, and a steady-state requirement for molecular biologists until a practical genetically engineered product is successfully marketed.

Industry and New Products for and from Agriculture

In Chapter 1, I described the organic pesticide revolution of the 1940s and 1950s. Will such a rapid revolution occur with the displacement of classical organic agrochemicals through biotechnology? I think not! The 1990s will continue to provide intense biotechnological research effort, but new molecularly engineered products therefrom will capture markets only slowly and gradually. Some areas of early promise are

1. areas of biotechnology other than the application of gene splicing for plants (fermentation products and processes, enzyme technology, etc.)
2. insecticidal, fungicidal, and herbicidal products from natural sources (antibiotics for medicine and agriculture, *Bacillus thuringiensis*, baculovirus, etc.)
3. molecular engineering, manipulation of prokaryotes for direct production of useful chemicals, insecticides, toxins, and various useful organic compounds (amino acids, fats, protein, and carbohydrates)

The practical manipulation of eukaryotes still requires more fundamental biological knowledge than currently exists. To be successful even for monogenic plant transformations, several hurdles must be surmounted. For example, the successful introduction of *B. thuringiensis* (BT) toxin into a plant can provide a useful product only if the problem of insect resistance is solved by some currently unknown strategy. In fact, if this hurdle is not overcome, not only will the plant variety fail, but natural products such as the BT organism or its self-generated peptide toxin could be rendered useless. For many genetically controlled properties mostly multigenic in character (e.g., various stress factors such as temperature, salt, and many disease and physiological tolerances), the pace for the introduction of new varieties through gene manipulation will be even slower.

Consequently, we can expect (1) a high demand in research and development (R&D) for scientists and engineers with biotechnological capabilities, (2) the demise of many of the new venture capital biotechnological companies, and (3) the consolidation of others with takeovers or working arrangements within the new venture capital companies and by and with the established agrochemical corporations. This consolidation is currently occurring.

R&D in the classical agrochemical areas will continue as expected for a mature industry rather than a new and expanding field. Great emphasis will be on solving the environmental and regulatory problems associated with classical products in general. A steady state of demand will exist for chemists trained in modern analytical techniques, computer applications, etc., with a slight decline in proportional numbers for synthesis chemists. New molecular structures will have to compete with low-cost, off-patent, toxicologically and environmentally acceptable old products alone and in mixtures.

It is becoming increasingly statistically improbable that large numbers of acceptable classic, simple pesticidal molecules will be discovered. The new product should preferably possess high activity at increasingly lower dosages to compete with old, off-patent, useful classic products and combinations thereof. There will be a continued need for classic reregistered agrochemicals, and development will center on combinations and strategies for their successful application, with emphasis on minimal unfavorable environmental impact.

The agrochemical industry will continue the trend of the 1980s with further consolidations, takeovers, and working arrangements within the industry (Table I). There will be an increase of acquisitions or working arrangements by the "giants" of the more promising and more successful venture capital companies. The trend toward the globalization of industry as a whole will affect the agrochemical industry. More and more foreign companies are acquiring American companies. (In the San Francisco Bay Area, five companies [Shell, Chevron, Dow,

Table I. Some Recent Takeovers or Mergers Involving the Agrochemical Industry

Company	Acquisition	Date
Avery	Acquired Uniroyal Chemical	1986
Bayer	Sold Helena Chemical Co. U.S. to Marubeni	1986
	Rhinechem Corp., Bayer AG's U.S. holding company, renamed Bayer USA Inc.	1986
BASF	Acquired Blazer herbicide (soybeans) from Rohm and Haas	1987
Chevron	Joint venture with Sumitomo Chemical to form Valent Corp.	1988
	Acquired PPG's agrochemical business	1988
	Sold back U.S. paraquat marketing rights to ICI	1986
Du Pont	Acquired Shell Agricultural Chemical (U.S.)	1986
ICI	Acquired Stauffer Chemical Group from Cheesebrough–Pond	1987
	Purchased paraquat marketing rights from Chevron	1986
Rhone–Poulenc	Acquired Union Carbide's agrochemical business	1986
	Acquired rights to CME 134 (IGR) from Celamerck	1986
	Acquired Stauffer Chemical's U.K. business	1985
	Acquired Niagara Chemicals (through May & Baker)	1984
	Purchased Diamond Shamrock Agrochemical (U.K.)	1983
	Acquired Mobil Ag Division	1981
Sandoz	Acquired VS Crop Protection (Velsicol)	1986
	Acquired Zoecon Corp.	1983
Schering	Name of FBC changed to Schering Agriculture	1986
	Acquired Nor-Am (Schering's U.S. operations)	1985
	Acquired controlling interest in the Boots–Hercules U.S. agrochemical business	1984
	Acquired FBC, a Fisons–Boots subsidiary (U.K.) agrochemical business of Upjohn	1984
Shell	Deutsche Shell acquired Celamerck	1987
	Sold U.S. agrochemical business to Du Pont	1986
Velsicol	Sold VS Crop Protection to Sandoz	1986
	Acquired by W. F. Farley (Chicago investor)	1985
Celamerck	Sold to Shell	1987
	Sold rights to CME 134 (IGR) to Rhone–Poulenc	1986
PPG	Agrochemical business sold to Chevron	1988
Rohm & Haas	Sold Blazer herbicide to BASF	1987
Stauffer	Sold to ICI	1987
	Sold to Chesebrough–Pond	

Table I.—*Continued*

Company	Acquisition	Date
Union Carbide	Sold agrochemical interest to Rhone–Poulenc	1986
Uniroyal	Sold chemicals business to Avery	1986
Elf Aquitaine	Acquired Pennwalt	1989
Dow	Joint venture with Eli Lilly's Elanco Division to form DowElanco	1989
Cyanamid	Acquired Celgene	1989
Ciba–Geigy	Acquired Dr Maag	1990

Stauffer, and Zoecon] were pursuing agrochemical discoveries. Currently two, both foreign-owned [ICI and Sandoz] remain.)

The ACS and the ACS Division of Agrochemicals

From what has preceded, in the future the ACS and the Division will struggle with multiple difficulties. I venture to project the following:

1. ACS membership will be older, and ACS will struggle to maintain that older membership. A portion of the younger people joining the ACS will exhibit some dissatisfaction and will continue to leave the ACS for organizations they perceive to be more relevant, particularly those relating to biotechnology.

2. Reflecting educational trends, the new ACS membership will show a rise in chemists of Asian ancestry and, in gender, a proportional increase of women.

3. Also as a reflection of educational trends, the ACS will struggle to recruit the new graduates who practice chemistry, but who derive from university biology departments. The ACS is regarded by these graduates as too large and too highly structured. They continue to prefer the smaller specialized biological societies. Perhaps the very real efforts of the ACS and the Division of Agrochemicals will ultimately counteract these perceptions. These efforts may in time be rewarded, but the current trend will not be reversed until a difficult-to-determine number of years has passed.

4. The Division of Agrochemicals will, at best, oscillate in number close to the present total membership for the near future.

U.S. Agriculture and World Agriculture

At the end of World War II, much of the developed world had been devastated and its peoples exhausted. For the undeveloped countries, a more complex situation existed, derived both directly from the war and particularly from the struggles against colonialism. The United States provided (with a motivation that was in part idealistic) food for much of the world. The value that transcended all others at that time was yield or productivity to assuage hunger. Furthermore, the Communist threat of absorbing most of Europe directly and of Asia, Africa, and Latin America less directly resulted in political and legislative support for American indigenous agricultural production. This food output fueled the Marshall Plan, private organizations such as the Ford Foundation, and programs of the U.S. government (AID and the Peace Corps). International Organizations such as the World Bank and UNIDO (United Nations Industrial Development Organization) all supplied aid relating to the increase of food supply. Not only food, but U.S. agricultural technology was freely exported to the Third World. It is not by chance that a Nobel prize was justly awarded to Norman E. Borlaug for the successful introduction of what was subsequently called the "Green Revolution". Many others, including this writer, contributed in a small way in the post-war decades toward this Green Revolution.

Modern agricultural technology is now shared throughout the world. Labor costs are much lower outside the United States, as are total input costs. Consequently, a real crisis exists in the competitive marketing of U.S. agricultural products. Furthermore, agriculture is a government-subsidized industry throughout the world. *For agricultural products, a world free-market economy simply does not exist.*

The future of agrochemicals and related industries depends in large part on the resolution of two opposing vectors. The first is those forces deriving from the public perception of science and, particularly of chemicals and their environmental impact. In the developed nations, both the European Community and the United States particularly, agricultural technology, especially chemical, is perceived with suspicion and some hostility. In the immediate future, these perceptions will have increasing legislative counterparts that in most cases are disadvantageous to U.S. agriculture and the agrochemical industries. The social, political, and economic consequences will be to restrict and limit production, registration, and use of agrochemicals.

The second force in the future of agrochemicals is the inexorable world population increases that necessitate increased production of food and fiber. The fundamental driving force toward environmental degradation today is population increase. Why are the rain forests of the world being decimated in Brazil, Central America, Malaysia, the Philippines, India, etc.? The answer is obvious: population pressure. Only two countries have taken official positions to control their populations. One is China, much of whose current economic and political woes derive from its billion people living in a denuded landscape. The other is India, which made an attempt at population control a decade ago, but had to abandon it because of opposition brought about by various unenlightened social forces.

Advances in science and technology (agricultural, industrial, and medical) are constantly resulting in improved human well-being and increased life expectancy, but these results exacerbate the problem of overpopulation. Aesthetics, the quality of life, and environmental consequences all point to the need for demographic rationality, that is, the need to limit this planet's human population.

Extrapolations resulting exclusively from population increase, manufacturing, production, and agrochemical sales over the past decades should point to a rosy future. However, nations without hard currency or new materials or manufactured products for exchange cannot buy American farm products or participate effectively in international trade. The population increase is greatest for such impoverished nations, for example, Bangladesh.

Sandoz, with which I was recently affiliated, is experiencing both for the United States and the world at large one of the best agrochemical sales and profit years (1990) in its history. Fur-

thermore, on the basis of the quarterly reports of larger and
smaller competitors, this highly successful pattern is the same
for most, if not all, of them. Agrochemical sales for 1988 increased 8.3% in France, 14.0% in the United Kingdom, and similarly for most of the European Community countries. Bayer,
Ciba-Geigy, ICI, Du Pont, Dow, Monsanto, etc., all reveal higher
sales and profits in 1989. Are the statistics for this year merely
artifacts, blips on a more general curve, for example, drought in
1988? No, but they cannot be used exclusively to predict the
future.

Agriculture will be affected by the various forces that I have
discussed. This view generates the hypothetical curve (Figure
1) that first shows a decline and then a rise that can be expected
for future trends. Examples of the trends that will follow this
curve are

- the volume of agrochemicals produced or sold
- the number of students in science or in agriculture
- the number of new classic agrochemicals registered
- the number of new biotechnological products registered
  and their sales volume

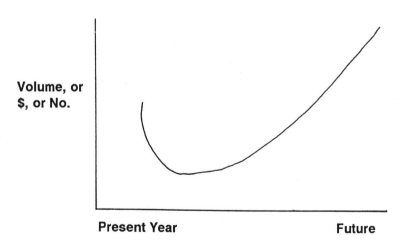

Figure 1. A generalized curve for future trends relating to agrochemicals.

Both the ordinate and the abscissa vary for each example, but in general we will experience considerable difficulties for the immediate future, with a recovery over the long run. In terms of this curve, our task is to limit the depth of the trough and its width on the time scale. For the ACS and for the agrochemical scientist this task combines our responsibilities on an educational level to reduce scientific illiteracy and prejudice within the general population and to provide scientific discovery and invention toward the improvement of agriculture in general.

Agricultural chemists also have the responsibility to (1) provide leadership to reduce contamination of the air, water, and land within our own country; and (2) provide agricultural chemicals derived from synthesis or biotechnology that are useful to agriculture and with minimal deleterious environmental impact.

The production of food is a noble endeavor, and we should take pride in our association with it. Agrochemicals, in the broadest sense that encompasses the new biotechnology as well, are an essential part of the technology that has greatly expanded our capacity for food production.

With one small additional phrase, designated by brackets, we can still enthusiastically endorse the social critic Jonathan Swift's statement of two centuries ago:

And he gavest for his opinion that whoever could make two ears of corn or two blades of grass [or provide new and better corn or grass] upon a spot of ground where only one grew before would deserve better of mankind and do more essential service to his country than the whole race of politicians.

Bear in mind that production statistics show that three to four ears of corn now grow in the United States where only one grew in the 1930s and 1940s. If scientists are to help to feed the world in the future, this productivity increase should provide the ecologically sound alternative to the slash and burn of the world's tropical forests, and this productivity can be obtained with cooperative efforts with minimal ecological harm to the cultivated

Indexes

Affiliation Index

Subject Index

A

Copy editing and indexing: Janet S. Dodd
Production: Margaret J. Brown and Paula M. Bérard
Cover design: Amy Hayes
Acquisition: Cheryl Shanks

Typeset by Techna Type Inc., York, PA, and
TypeWorks Plus Inc., Silver Spring, MD
Books printed and bound by Maple Press, York, PA

The paper used in this publication meets the
minimum requirements of American National
Standard for Information Sciences—Permanence
of Paper for Printed Library Materials,
ANSI Z39.48–1984 ⊚

Bestsellers from ACS Books

The ACS Style Guide: A Manual for Authors and Editors
Edited by Janet S. Dodd
264 pp; clothbound, ISBN 0–8412–0917–0; paperback, ISBN 0–8412–0943–X

Chemical Activities and *Chemical Activities: Teacher Edition*
By Christie L. Borgford and Lee R. Summerlin
330 pp; spiralbound, ISBN 0–8412–1417–4; teacher ed. ISBN 0–8412–1416–6

Chemical Demonstrations: A Sourcebook for Teachers,
Volumes 1 and 2, Second Edition
Volume 1 by Lee R. Summerlin and James L. Ealy, Jr.;
Vol. 1, 198 pp; spiralbound, ISBN 0–8412–1481–6;
Volume 2 by Lee R. Summerlin, Christie L. Borgford, and Julie B. Ealy
Vol. 2, 234 pp; spiralbound, ISBN 0–8412–1535–9

Writing the Laboratory Notebook
By Howard M. Kanare
145 pp; clothbound, ISBN 0–8412–0906–5; paperback, ISBN 0–8412–0933–2

Developing a Chemical Hygiene Plan
By Jay A. Young, Warren K. Kingsley, and George H. Wahl, Jr.
paperback, ISBN 0–8412–1876–5

Biotechnology and Materials Science: Chemistry for the Future
Edited by Mary L. Good (Jacqueline K. Barton, Associate Editor)
135 pp; clothbound, ISBN 0–8412–1472–7; paperback, ISBN 0–8412–1473–5

Personal Computers for Scientists: A Byte at a Time
By Glenn I. Ouchi
276 pp; clothbound, ISBN 0–8412–1000–4; paperback, ISBN 0–8412–1001–2

Polymeric Materials: Chemistry for the Future
By Joseph Alper and Gordon L. Nelson
110 pp; clothbound, ISBN 0–8412–1622–3; paperback, ISBN 0–8412–1613–4

Chemical Structure Software for Personal Computers
Edited by Daniel E. Meyer, Wendy A. Warr, Richard A. Love
107 pp; clothbound, ISBN 0–8412–1538–3; paperback, ISBN 0–8412–1539–1

For further information and a free catalog of ACS books, contact:
American Chemical Society
Distribution Office, Department 225
1155 16th Street, NW, Washington, DC 20036
Telephone 800–227–5558

Highlights from ACS Books

Silent Spring Revisited
Edited by Gino J. Marco, Robert M. Hollingworth, and William Durham
214 pp; clothbound, ISBN 0–8412–0980–4; paperback, ISBN 0–8412–0981–2

Chemistry and Crime: From Sherlock Holmes to Today's Courtroom
Edited by Samuel M. Gerber
135 pp; clothbound, ISBN 0–8412–0784–4; paperback, ISBN 0–8412–0785–2

From Caveman to Chemist: Circumstances and Achievements
By Hugh W. Salzberg
300 pp; clothbound, ISBN 0–8412–1786–6; paperback, ISBN 0–8412–1787–4

The Green Flame: Surviving Government Secrecy
By Andrew Dequasie
300 pp; clothbound, ISBN 0–8412–1857–9

Trends in Chemical Consulting
Edited by Charles S. Sodano and David M. Sturmer
165 pp; paperback ISBN 0–8412–2106–5

The Language of Biotechnology: A Dictionary of Terms
By John M. Walker and Michael Cox
254 pp; clothbound, ISBN 0–8412–1489–1; paperback, ISBN 0–8412–1490–5

The Basics of Technical Communicating
By B. Edward Cain
198 pp; clothbound, ISBN 0–8412–1451–4; paperback, ISBN 0–8412–1452–2

Phosphorus Chemistry in Everyday Living
By Arthur D. F. Toy and Edward N. Walsh
362 pp; clothbound, ISBN 0–8412–1002–0

Steroids Made It Possible
By Carl Djerassi
205 pp; clothbound, ISBN 0–8412–1773–4

For further information and a free catalog of ACS books, contact:
American Chemical Society
Distribution Office, Department 225
1155 16th Street, NW, Washington, DC 20036
Telephone 800–227–5558